Cracking the Data Science Interview

Unlock insider tips from industry experts to master the data science field

Leondra R. Gonzalez

Aaren Stubberfield

Cracking the Data Science Interview

Group Product Manager: Niranjan Naikwadi
Publishing Product Manager: Nitin Nainani
Senior Editor: Hayden Edwards
Technical Editor: Simran Haresh Udasi
Copy Editor: Safis Editing
Project Coordinator: Aishwarya Mohan
Proofreader: Safis Editing
Indexer: Rekha Nair
Production Designer: Prashant Ghare
Marketing Coordinators: Vinishka Kalra

First published: March 2024
Production reference: 1160224

Published by Packt Publishing Ltd.
Grosvenor House
11 St Paul's Square
Birmingham
B3 1RB

ISBN 978-1-80512-050-6

www.packtpub.com

Foreword

The data science landscape is ever-evolving and has been that way since its conception. Though it is a rewarding field with many opportunities, navigating it can be a challenge, especially when you're just getting started.

During my career, I have found that various companies can interpret data science differently depending on their business needs or understanding of data science. When I first began my data science journey in 2015, I was employed as a health data analyst with a start-up. It was there that I was exposed to data science, as my role was not purely data analytics or data science, but a mixture somewhere in between. I wanted to continue learning and advancing, but I did not know where to focus my energy to gain the information needed to thrive in this field. So, I curated a list of lessons I needed to learn in order to be competent enough to enter and advance in the field. I learned Python, data science with Python, R programming, linear algebra, and calculus, and as time went on, it became more and more daunting, the list of lessons becoming even longer than what was required for a graduate degree. Unfortunately, even after all of my hard work, during interviews, I found there were still concepts that I was unaware of. This has been the issue that I, as well as others, have noted with this field – there is so much information, but it can be unclear where to begin and what information is necessary to know.

On top of this, the data science interview is universally dreaded and challenging for various reasons that I have already alluded to. For instance, candidates are usually unsure of what that particular company considers data science. Plus, take-home assignments can take hours to complete – and once that time has been invested in completing the assignment, the company may choose to not offer feedback or, even worse, disappear completely when they've decided they aren't interested. After experiencing this devastating outcome more than once, I became highly selective in what companies I chose to do a take-home assignment for. Many companies had a habit of immediately asking candidates to complete a take-home assignment before an interview, which I have learned rarely works in the candidate's favor.

This book will address and outline the concepts that are necessary to begin or progress in a data science role. Because this field is ever-evolving, our understanding of concepts will continue as well, however this book can be used as a reference for those that are experienced in the field, or for those that are in data science adjacent roles and want to keep their knowledge current. This book will include imperative information so that candidates can be successful during a data science interview, as well as removing some of the guesswork in what companies are expecting.

It is widely accepted that data science candidates have an online portfolio to showcase their talent and application of knowledge – for this reason, there is information on how to build a portfolio and create a resume that will get you noticed. Salary and benefits negotiation is also outlined to streamline the process for you – a process many of us had to learn completely uninformed in the past, is now disseminated for the benefit of others.

We are certain that you will find this book helpful in your data science journey. Cheers!

Angela Baltes, PhD

Data Scientist, UnitedHealth Group

Contributors

About the authors

Leondra R. Gonzalez is a senior data and applied scientist at Microsoft with a decade of experience in data science, analytics, and corporate strategy. In addition to her work as a data scientist, Leondra has led teams in the entertainment, media, and advertising space to produce advanced e-commerce models for top brands, including NBC Peacock, First Aid Beauty, Procter & Gamble, HBO Max, Toyota, Whirlpool, and Tubi.

Academically, Leondra graduated from Carnegie Mellon University's Heinz College of Information Systems Management with a master's in entertainment industry management, with a focus on business analytics; Quantic School of Business and Technology with an MBA, including a specialization in statistics; and Otterbein University with a bachelor's in music and business. Leondra is currently pursuing a PhD in information technology with a specialization in artificial intelligence at the University of the Cumberlands, and she has researched deep learning architectures as a PhD computer science apprentice at Google.

To my loving husband, Chris, my parents, my sister, and my unborn son who kicked my bump every day while writing this book.

Aaren Stubberfield is a senior data scientist for Microsoft's digital advertising business and the author of three popular courses on DataCamp. He graduated with an MS in predictive analytics and has over 10 years of experience in various data science and analytical roles, focused on finding insights for business-related questions.

With his experience, he has led numerous teams of data scientists and has been instrumental in the successful completion of many projects. Aaren's technical skills include the use of AI, like LLMs, Python, and various other tools necessary for the execution of data science projects.

I want to thank the people who have been close to me and supported me, especially my wife, Pam, and my family.

About the reviewer

Vishal Kumar, a seasoned data scientist, has over seven years of experience with a premium credit card company, where he has made indelible contributions to the realms of AI and ML. He has a master's degree in statistics from Delhi University.

Throughout his career, he has garnered a plethora of accolades, stemming from his adeptness in constructing cutting-edge decision science tools that have steered various organizations' success. His commitment to continuous learning is evidenced by his embrace of new technologies, such as generative AI, to stay at the forefront of the ever-evolving data science landscape.

Beyond his professional pursuits, his creativity extends into his personal life, as he likes to paint and play ukulele.

Table of Contents

2

Finding a Job in Data Science   25

Part 2: Manipulating and Managing Data

3

Programming with Python   51

4

Visualizing Data and Data Storytelling   87

5

Querying Databases with SQL   121

6

Scripting with Shell and Bash Commands in Linux   161

7

Using Git for Version Control   181

Part 3: Exploring Artificial Intelligence

8

Mining Data with Probability and Statistics   197

9

Understanding Feature Engineering and Preparing Data for Modeling   229

10

Mastering Machine Learning Concepts   259

11

Building Networks with Deep Learning   295

12

Implementing Machine Learning Solutions with MLOps   323

Part 4: Getting the Job

13

Mastering the Interview Rounds   341

14

Preface

In today's dynamic technological landscape, the demand for skilled professionals in **artificial intelligence (AI)** and data science roles has surged, and the data science job market is increasingly saturated by various levels of data science and AI employees. This book is a comprehensive guide, crafted to equip both aspiring and seasoned individuals with the essential tools and knowledge required to navigate the intricacies of data science interviews. Whether you're stepping into the AI realm for the first time or aiming to elevate your expertise, this book offers a holistic approach to mastering the fundamental and cutting-edge facets of the field.

The chapters within this book span a wide spectrum of critical subjects, from programming with Python and SQL to statistical analysis, pre-modeling and data cleaning concepts, **machine learning (ML)**, deep learning, **Large Language Models (LLMs)**, and generative AI. We aim to provide a comprehensive review and update on the foundational concepts while also delving into the latest advancements. In an era marked by the disruptive potential of language models and generative AI, it's imperative to continually hone your skills. This book serves as a compass, guiding you through the intricacies of these transformative technologies, ensuring you're poised to tackle the challenges and harness the opportunities they present.

Moreover, beyond technical prowess, we delve into the art of interviewing for AI roles, offering guidance on how to ace interviews and negotiate compensation effectively. Additionally, crafting a standout résumé tailored for data science roles is a crucial step, and our guide offers insights into writing compelling résumés that capture attention in a competitive job market. As AI reshapes industries and innovation accelerates, now is the ideal time to embark on or advance in your data science journey. We invite you to dive into this comprehensive resource and embark on your path to mastering the dynamic world of data science and AI.

Who this book is for

If you are a seasoned or young professional who needs to brush up on your technical skills, or you are looking to break into the exciting world of the data science industry, then this book is for you.

What this book covers

In *Chapter 1*, *Exploring the Modern Data Science Landscape*, we begin our journey with a brief but valuable overview of the contemporary landscape of data science and AI.

In *Chapter 2*, *Finding a Job in Data Science*, we will introduce data science roles and their various categories.

In *Chapter 3, Programming with Python*, you will familiarize yourself with the most common and useful tasks and operations in the Python language.

In *Chapter 4, Visualizing Data and Storytelling*, you will learn techniques for telling engaging data stories.

In *Chapter 5, Querying Databases with SQL*, you will dive into the world of databases, understanding their design and how to query them to acquire data.

In *Chapter 6, Scripting with Bash and Shell Commands in Linux*, you will boost your operating system skills with the power of bash and shell commands, enabling you to interface with multiple technologies either locally or in the cloud.

In *Chapter 7, Using Git for Version Control*, we explore the most useful commands in Git for project collaboration and reproducibility.

In *Chapter 8, Mining Data with Probability and Statistics*, you will understand some of the most relevant topics in probability and statistics that serve as the foundation for many ML models and assumptions.

In *Chapter 9, Understanding Feature Engineering and Preparing Data for Modeling*, you will use your understanding of descriptive statistics to create clean, "machine-legible" datasets.

In *Chapter 10, Mastering Machine Learning Concepts*, you will learn about the most used ML algorithms, their assumptions, how they work, and how to best evaluate their performance.

In *Chapter 11, Building Networks with Deep Learning*, we take a step further into building and evaluating neural networks in various applications while also touching base on the latest advancements in AI.

In *Chapter 12, Implementing Machine Learning Solutions with MLOps*, we will review the data science process, tools, and strategies to effectively design and implement an end-to-end ML solution.

In *Chapter 13, Mastering the Interview Rounds*, you will learn the best techniques to successfully bypass technical and non-technical factors at every stage of the interview process.

In *Chapter 14, Negotiating Compensation*, you will learn to optimize your earning potential.

To get the most out of this book

To get the most out of this book, you should have a basic knowledge of Python, SQL, and statistics. However, you will also benefit from this book if you have familiarity with other analytical languages, such as R. By brushing up on critical data science concepts such as SQL, Git, statistics, and deep learning, you'll be well-equipped to crack through the interview process.

Software/hardware covered in the book	Operating system requirements
Python 3.12	Windows, macOS, or Linux
Bash	Linux
Jupyter Notebooks	Windows, macOS, or Linux

Conventions used

There are a number of text conventions used throughout this book.

`Code in text`: Indicates code words in text, database table names, folder names, filenames, file extensions, pathnames, dummy URLs, user input, and Twitter handles. Here is an example: The `split()` method can be used to split s into individual words: `words = s.split()`.

A block of code is set as follows:

```
x = 5
print(type(x)) # <class 'int'>
```

Bold: Indicates a new term, an important word, or words that you see on screen. For instance, words in menus or dialog boxes appear in **bold**. Here is an example: "The increased computing power and the development of advanced algorithms, especially in **machine learning** (**ML**) and **deep learning** (**DL**), have made it possible to efficiently process and analyze massive amounts of data."

> **Tips or important notes**
> Appear like this.

Special Note

The prevalence of accessible AI technology has exploded over the past few months, particularly over the course of writing this book. We encourage our readers to utilize AI during their educational journey, leveraging tools such as Chat GPT to test your newly acquired skills. Long gone are the days where you browse StackOverFlow for hours for your specific inquiry. Now, the power of asking for help is right at your fingertips.

Even we, the authors of this book, leveraged generative AI to aid in minor editorial tasks and creating code examples. However, rest assured that humans wrote the content and laid out what is covered in the book! In this new era, we just wanted to make our readers aware of how we used the tool.

Get in touch

Feedback from our readers is always welcome.

General feedback: If you have questions about any aspect of this book, email us at customercare@packtpub.com and mention the book title in the subject of your message.

Errata: Although we have taken every care to ensure the accuracy of our content, mistakes do happen. If you have found a mistake in this book, we would be grateful if you would report this to us. Please visit www.packtpub.com/support/errata and fill in the form.

Piracy: If you come across any illegal copies of our works in any form on the internet, we would be grateful if you would provide us with the location address or website name. Please contact us at copyright@packtpub.com with a link to the material.

If you are interested in becoming an author: If there is a topic that you have expertise in and you are interested in either writing or contributing to a book, please visit authors.packtpub.com.

Share Your Thoughts

Once you've read *Cracking the Data Science Interview*, we'd love to hear your thoughts! Scan the QR code below to go straight to the Amazon review page for this book and share your feedback.

https://packt.link/r/1-805-12050-6

Your review is important to us and the tech community and will help us make sure we're delivering excellent quality content.

Download a free PDF copy of this book

Thanks for purchasing this book!

Do you like to read on the go but are unable to carry your print books everywhere?

Is your eBook purchase not compatible with the device of your choice?

Don't worry, now with every Packt book you get a DRM-free PDF version of that book at no cost.

Read anywhere, any place, on any device. Search, copy, and paste code from your favorite technical books directly into your application.

The perks don't stop there, you can get exclusive access to discounts, newsletters, and great free content in your inbox daily

Follow these simple steps to get the benefits:

1. Scan the QR code or visit the link below

https://packt.link/free-ebook/978-1-80512-050-6

2. Submit your proof of purchase
3. That's it! We'll send your free PDF and other benefits to your email directly

Part 1:
Breaking into the
Data Science Field

In the first part of this book, you will learn about the data science profession as it exists in the modern day, and how this relates to your endeavors in the field. This will serve as an introduction to various career paths and help to set expectations in terms of the skills and competencies required to be successful.

This part includes the following chapters:

- *Chapter 1, Exploring Today's Modern Data Science Landscape*
- *Chapter 2, Finding a Job in Data Science*

1

Exploring Today's Modern Data Science Landscape

If you've picked up this book, chances are that you've already heard of data science. It's arguably one of the fastest-growing, most discussed professions within the tech and STEM space, all while maintaining its relative edge and mystique. That is, many people have heard of data scientists, but very few know what they do, how a data scientist produces value, or how to *break into* the field from scratch.

In this chapter, we will verify the definition of data science with a practical description. Then, we will discuss what most data science jobs entail, while spending some time describing the distinction between different flavors of data science. We'll then dive into the various paths into data science and what makes it so challenging to land your first job. We'll finish the chapter with an overview of the non-negotiable competencies expected of data scientists.

By the end of this chapter, you will have a firm understanding of the modern data scientist, the various paths to getting the job, and what to expect in your journey to becoming one.

With this gentle introduction, you'll have a better understanding of the job of a data scientist, which path to becoming a data scientist best fits your journey, the barriers to expect in your journey, and which skills you should master.

In this chapter, we will cover the following topics:

- What is data science?
- Exploring the data science process
- Dissecting the flavors of data science
- Reviewing career paths in data science
- Tacking the experience bottleneck
- Understanding expected skills and competencies
- Exploring the evolution of data science

What is data science?

To begin, let's offer a definition of data science. According to *Wikipedia*, **data science** "*is an interdisciplinary academic field that uses statistics, scientific computing, scientific methods, processes, algorithms, and systems to extract or extrapolate knowledge and insights from noisy, structured, and unstructured data*"[1]. It encompasses various techniques, procedures, and tools to process, analyze, and visualize data, enabling businesses and organizations to make data-driven decisions and predictions. The primary goal of data science is to identify patterns, relationships, and trends within data to support decision-making and create actionable insights.

You are not alone in your interest in data science – it was called by the *Harvard Business Review* one of the sexiest jobs in the 21st century [2], and stories of data scientists earning enormous salaries in the six-figure range are not uncommon. Data scientists are often looked at as oracles within an organization, answering complex business questions such as, "If we increase our offering to this group of customers, can we increase our revenues?" or "What are the common causes of customer churn?"

Within organizations, the demand for the skills of data scientists has continued to grow. The U.S. Bureau of Labor Statistics estimated that in 2022, the number of jobs for data scientists will increase by roughly 36% over the next 10 years [3]. This growth in the demand for data scientists is being fuelled by several factors, which are shown here:

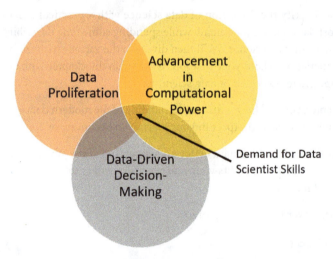

Figure 1.1: Reasons for the increased demand for data scientists

The first is the proliferation of data. The exponential growth of data generated by digital devices, social media, and various other sources has made it essential for organizations to harness this data for decision-making and innovation. This data growth is expected to continue in the future, with the **International Data Corporation** (**IDC**) expecting that by 2025, we will generate 175 zettabytes of data annually [4]. That is a staggering amount of data!

Organizations want to take advantage of this explosion in data availability to generate insights for decision-making. As the world becomes more interconnected and complex, the need for evidence-based decision-making has grown, leading to an increased demand for skilled data scientists who can transform data into actionable insights. Organizations and businesses increasingly rely on data-driven insights to gain a competitive edge in the market, optimize operations, and improve customer experiences.

Finally, transforming data into insights couldn't be accomplished without advancements in computational power and the advancement of tools and platforms. The increased computing power and the development of advanced algorithms, especially in **machine learning** (**ML**) and **deep learning** (**DL**), have made it possible to efficiently process and analyze massive amounts of data. In addition, the development of open source tools, libraries, and platforms has made data science more accessible to a broader audience, fostering the growth of the profession.

Hence, data science is still an evolving field that is only expected to grow in parallel with computational and technological advancements (such as generative AI). Furthermore, as companies continue to embrace the digital age with an increased interest in maximizing their utility of data and capitalizing on its underlying insights for a competitive advantage, the demand for data scientists will also expand.

However, although data science is often regarded and described as a monolithic function, you'll soon learn that it's a multi-faceted discipline that often varies by team, department, or even company. Naturally, the data scientist job profile is also an ever-evolving description, but we will cover all our bases for the most common tasks.

Exploring the data science process

Performing data science work is often an iterative process, where the data scientist needs to return to earlier steps if they run into challenges. There are many ways to categorize the data science process, but it often includes:

- Data collection
- Data exploration
- Data modeling
- Model evaluation
- Model deployment and monitoring

Let's briefly touch on each step and discuss what's expected of the data scientist during them.

Data collection

Data collection and preprocessing involves gathering data from various sources (such as databases, APIs, and web scraping), then cleaning and transforming the data to prepare it for analysis. This step involves dealing with missing, inconsistent, or noisy data and converting it into a structured format.

Depending on the organization, a team of data engineers support this step of the data science process; however, it is common for the data scientist to manage this process as well. This requires them to have intimate knowledge of the data sources and the ability to write **Structured Query Language (SQL)** queries, code that can query databases, or custom tools such as web scrapers to gather the needed data.

Data exploration

Data exploration involves conducting **exploratory data analysis (EDA)** to better understand the data, detect anomalies, and identify relationships between variables. The key to this step is to look for correlations and understand the distribution of the data. This involves using descriptive statistics and visualization techniques to summarize the data and gain insights; therefore, the data scientist should be able to use summary statistics, program descriptive visualizations, or utilize reporting tools such as Power BI or Tableau to create robust charts.

Data modeling

Using what was learned in the data exploration step, **data modeling** is the step when the data scientist builds their predictive or descriptive models using ML and statistical techniques that identify patterns and relationships in the data. Here, the data scientist selects the appropriate algorithms, trains the models on historical data, and validates their performance.

Model evaluation

Model evaluation and optimization involves assessing the performance of models using metrics such as accuracy, RMSE, precision, recall, AUC, or F1 scores. Based on these evaluations, data scientists may refine the models or try alternative algorithms to improve their performance. Understanding the underlying reasons behind a model's predictions is crucial for building trust in its results and ensuring that it aligns with the domain knowledge. Therefore, the data scientist must be sure the model solves the organizational/business goal. Here, the data scientist needs to be able to communicate their findings to possible technical and non-technical individuals.

Model deployment and monitoring

Model deployment and monitoring involves implementing the models in real-world applications, monitoring their performance, and maintaining them to ensure their continued accuracy and relevance. For example, the data scientist might work with a data engineering team or use tools such as containers to implement the model. Once deployed, the data scientist may also need to develop dashboards to monitor the model's performance over time and flag stakeholders if it goes outside the expected performance range.

As you can see, data science is a profession that incorporates many data-related tasks – particularly those that involve the acquisition, prepping, and delivery of data in one format or another. While data modeling makes up most of the glitz and glamour associated with the job, it is really everything else that takes up roughly 80% of the gig. This does not include non-data-related tasks, such as interfacing with stakeholders, gathering requirements, debugging software, checking emails, and research. However, those tasks are not necessarily unique to data scientists.

Now that you understand the common tasks associated with the job, let's explore the different types or flavors of data science.

Dissecting the flavors of data science

Now that we have defined some of the critical aspects of the role of a data scientist, it is clear that the role often covers many different skills. Data scientists are frequently asked to perform a variety of data-related tasks, including designing database tables to collect data, programming ML algorithms, understanding statistics, and creating stunning visuals to help explain interesting findings to others, but it is difficult for any single person to master all of these skill areas.

Therefore, we often see data scientists who are particularly skilled in one or two areas and have basic competencies in the others. Their talents could be considered T-shaped, where they are proficient across many areas such as the horizontal line of a T, while they have deep knowledge and expertise in a few areas such as the vertical portion of the letter:

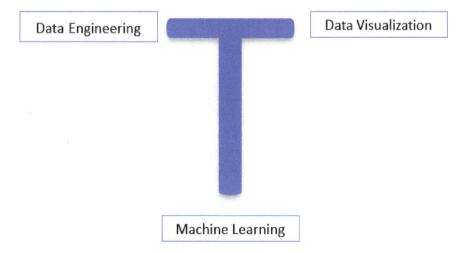

Figure 1.2: Example of the 'T of Competencies'

While this example shows an example of someone who is adequate in data engineering and visualization principles but exceptional in ML, you can expect to see every possible combination of skills among data scientists. These competencies are often aligned with a person's unique experiences or interests. Perhaps they were a statistics major and took a liking to ML, or perhaps they're a former **business intelligence** (**BI**) engineer with considerable experience in data **extraction, transformation, and loading** (**ETL**), allowing them to grasp data engineering concepts much faster.

Whatever the reason, it's natural for someone to grasp some concepts better than others. This is important to remember as you navigate this book. While you are not expected to specialize in every facet of data science, you are expected to master the fundamentals. However, you will almost certainly discover your *T of Competencies* – a trinity of top skill sets that will solidify your identity in the data science space.

While there are countless combinations of skill proficiencies, let's review some of the most common that you will encounter:

- The data engineer
- The dashboarding and visual specialist
- The ML specialist
- The domain expert

Let's take a look at these now.

Data engineer

As we discussed earlier, data engineering is a crucial aspect of the data science process that involves data collection, storage, processing, and management. It focuses on designing, developing, and maintaining scalable data infrastructure, ensuring the availability of high-quality data for analysis and modeling. **Data engineers** are most known for their oversight of the ETL process of data pipelines. On some data scientist teams, especially within smaller organizations, the data engineering responsibilities sit within the data science team. Therefore, the data scientist specializing in this area can help support team projects with data collection and storage, understanding the needs of the ML process, such as structuring the data so that it can be fed efficiently to a DL algorithm.

Data engineers have a wealth of tools to choose from. It is not expected for any single data engineer to know all of these technologies, especially at the same level of competencies. In fact, the more senior the engineer, the more competent they are in their tools of choice. Furthermore, this is not a comprehensive list. However, you can expect to see the following on data engineer resumes:

- *Programming languages*: Python, SQL, Scala, R, C++
- *Data storage*: Relational databases (for example, MySQL, PostgreSQL, Oracle), NoSQL databases (for example, MongoDB, Cassandra, DynamoDB), data warehouses (for example, Snowflake,

Redshift, BigQuery), distributed filesystems (for example, Hadoop Distributed File System (HDFS), Apache Cassandra)

- *Data processing and analysis*: Apache Spark, Apache Flink, Apache Storm, Apache Beam, MapReduce, Hadoop, Hive, Apache Kafka, Amazon Kinesis

- *Data integration and ETL*: Apache NiFi, Talend, Apache Airflow, AWS Glue, Google Cloud Dataflow, dbt

- *Data version control and collaboration*: Git, GitHub, GitLab, Bitbucket, Azure DevOps

- *Data visualization and BI*: Tableau, Power BI, Looker, QlikView, Domo

- *Cloud platforms and infrastructure*: Microsoft Azure, Google Cloud Platform (GCP), Amazon Web Services (AWS)

- *Containers*: Docker, Kubernetes

Dashboarding and visual specialist

Data visualization is the graphical representation of data and information using visual elements such as charts, graphs, and maps. It enables stakeholders to understand complex patterns, trends, and relationships in data, allowing for more informed decision-making. Data visualization helps simplify complex data and present it in an easily digestible format, identify patterns, trends, and correlations in data, support data-driven decision-making, and communicate insights and findings effectively to a broad audience. Combining data visualizations with a compelling narrative can become a powerful motivator to drive organizational actions. Many news organizations hire phenomenal data scientists specializing in data visualization to communicate complex information to their audience.

Dashboarding and visual specialists have different designations depending on the organization, but some of the most common names you'll hear include BI engineer, data analyst, data visualization expert, data storyteller, and many others. They are commonly individuals with a strong background in descriptive statistics, data storytelling, and developing **key performance indicators** (also known as **KPIs**). The most common tools you will see used by dashboarding and visual specialists include:

- *Programming languages*: Python, SQL, R, JavaScript

- *Data storage*: Relational databases (for example, MySQL, PostgreSQL, Oracle), NoSQL databases (for example, MongoDB, Cassandra, DynamoDB), data warehouses (for example, Snowflake, Redshift, BigQuery)

- *Frameworks*: Dask, Plotly, ggplot2, Shiny, Matplotlib, Seaborn, DB.js

- *Data visualization and BI*: Tableau, Power BI, Looker, QlikView, Domo, Funnel, Excel

- *Cloud platforms and infrastructure*: Microsoft Azure, GCP, AWS

ML specialist

When most people think about data scientists, they think about someone who designs and implements ML algorithms. **ML specialists and engineers** utilize computers to learn and improve from experience without explicit programming by developing algorithms and models to analyze data, identify patterns, and make predictions or decisions based on those patterns. They play a critical role in building intelligent applications and systems. ML specialists have a strong sense of which learning algorithms to use and how to adjust their parameters to achieve the best performance.

As a result, they have a strong propensity toward research to stay current on the latest methods of quantitative problem-solving and are specifically skilled in ML development, deployment, and maintenance tasks. They have a robust toolset as they are highly proficient in software development principles. While it certainly isn't a rule, many ML specialists tend to have a strong background in statistics, operations research, computer science, and/or information systems. Tools used by ML specialists might include:

- *Programming languages*: Python, SQL, R, Java, C++
- *Frameworks*: TensorFlow, Keras, scikit-learn, PyTorch, H2O, Hugging Face
- *Data storage*: Relational databases (for example, MySQL, PostgreSQL, Oracle), NoSQL databases (for example, MongoDB, Cassandra, DynamoDB), data warehouses (for example, Snowflake, Redshift, BigQuery), distributed filesystems (for example, HDFS, Apache Cassandra)
- *Data processing and analysis*: Apache Spark, Apache Flink, Apache Storm, Apache Beam, MapReduce, Apache Kafka
- *Data integration and ETL*: Apache NiFi, Talend, Apache Airflow, AWS Glue, Google Cloud Dataflow
- *Data version control and collaboration*: Git, GitHub, GitLab, Bitbucket
- *Cloud platforms and infrastructure*: Microsoft Azure, GCP, AWS
- *Deployment*: Docker, Kubernetes, Flask

Domain expert

Domain experts are data scientists with in-depth knowledge and expertise in specific domains within the industry or field; for example, someone who has gained much knowledge and expertise working on **computer vision** (**CV**) or **natural language** (**NL**) problems. They leverage their domain knowledge to develop custom ML models and data analysis techniques tailored to their domain's unique challenges and requirements. However, there are also non-technical domain experts who gained a deep familiarity with a particular industry or business problem given their professional history. For example, someone with a background in digital marketing may have an edge for a data science role that requires an understanding of media mix modeling or data-driven attribution, whereas someone with aviation experience may have an advantage in route optimization models.

Because domain experts tend to carry domain-specific expertise, they often are already familiar with the tools of their specific industry. For example, a digital marketing professional is bound to have some experience with a myriad of MarTech platforms, including Google Analytics, Adobe Analytics, HubSpot, and more.

These are just some of the flavors or different areas to specialize in within data science. You will not need to be an expert in all of these areas, but you will need to show some level of competency and willingness to grow in all of these areas. Often when working on data science projects, you will gravitate to one of these areas out of necessity or passion; gaining practical experience will be key here and strengthen your candidacy for a role where the hiring manager is looking for someone with that skill set.

If you haven't noticed, many of these data science flavors are the consequence of one's prior experience, either in tech or otherwise. For example, a software engineer may be well suited to transition into ML or data engineering, while a data analyst may find an easier time transitioning to data engineer or BI engineer. As you've seen, there is a considerable overlap in skills, tools, and tasks with all flavors of data science.

This brings us to the paths to data science. You may have already envisioned where you fit into the equation given some of the prior descriptions. Let's take the time to explicitly discuss some common paths to the data science profession.

Reviewing career paths in data science

The field of data science is rapidly evolving, drawing professionals from various backgrounds and disciplines. This dynamic landscape has given rise to a multitude of career paths in data science, each bringing their unique perspectives, skills, and experiences to the table. In this section, we will explore three primary types of data scientists: the traditionalist, the domain expert, and the off-the-beaten path-er. Does one of these career paths best fit you?

The traditionalist

The traditionalist data scientist has followed a more conventional educational path toward data science. They typically possess a strong background in computer science or mathematics, often with a minor in the other. Other common majors include operations research, statistics, physics, and engineering. These individuals often go on to earn an advanced degree in these fields, including a master's degree or even a Ph.D. Their rigorous academic training equips them with a deep understanding of statistical methodologies, programming languages, and advanced algorithms.

The traditionalist data scientist has a comprehensive understanding of the underlying mathematical and statistical principles that govern the field of data science. They are well-versed in probability theory, linear algebra, calculus, and optimization techniques, which form the basis for many ML algorithms and statistical modeling. This theoretical foundation enables them to grasp the nuances of various methods and research the most appropriate approach for a given problem.

Equipped with a background in computer science, traditionalists are adept at programming languages commonly used in data science, such as Python and R. Their programming skills allow them to manipulate data, implement ML algorithms, and develop custom solutions tailored to specific problems. Furthermore, they are skilled in using specialized libraries and frameworks, such as TensorFlow, PyTorch, and scikit-learn, to expedite the development of data science projects.

In brief, the traditionalist data scientist is characterized by their strong STEM academic background, comprehensive understanding of statistical principles, and proficiency in programming and data manipulation. If your background is traditionalist, we suggest positioning yourself in job interviews as someone with deep expertise in ML. In addition, highlight any research experience you have.

Domain expert

Domain expert data scientists are professionals who initially started their careers in a specific industry, such as marketing, finance, healthcare, or supply chain, before branching out into data science. With a strong understanding of their domain, these individuals have gradually acquired data analysis and programming skills to supplement their expertise (for example, a company controller uses domain expertise and knowledge to develop an ML algorithm that flags fraudulent transactions). Domain experts possess a unique ability to leverage their domain knowledge to uncover relevant insights from data, enabling organizations to make data-driven decisions that drive growth and efficiency.

Domain experts have a comprehensive understanding of the intricacies and nuances of their industry, making them invaluable assets in data-driven projects. Their knowledge of industry-specific challenges, trends, and best practices enables them to identify critical business problems and frame data-driven solutions that are relevant and impactful. Armed with extensive domain knowledge and analytical skills, domain expert data scientists excel at developing solutions tailored to their industry. In addition, they have a keen ability to translate business questions into data-driven hypotheses and use their understanding of the sector's unique characteristics to guide their analysis. This targeted approach allows them to generate insights that directly address the needs and priorities of their industry.

Additionally, domain experts are well versed in the analytical tools and software commonly used in their respective fields. These specialized tools, which may include industry-specific data platforms, visualization software, or ML frameworks, allow them to efficiently process and analyze data unique to their domain. Their expertise with these tools enables them to deliver insights more quickly and effectively than their counterparts who lack industry-specific knowledge.

Finally, one of the critical strengths of domain expert data scientists is their ability to communicate complex data insights to non-technical stakeholders within their industry. In addition, they understand the context and terminology of their domain, enabling them to present findings in a manner that resonates with their business partners. This skill is critical for driving data-driven decision-making and ensuring that the value of their work is recognized and understood by their organization.

In summary, if you have specialized knowledge of the field you are interviewing for, we suggest positioning yourself as a domain expert data scientist. Highlight your deep understanding of the industry and their challenges, enabling you to deliver targeted and impactful data-driven solutions. Additionally, highlight that you can communicate complex insights effectively using industry terminology. Your domain knowledge and data science techniques will make you a valuable asset to any organization in their field.

Off-the-beaten path-er

The off-the-beaten path-er data scientist is an individual who has ventured into data science from what's deemed as a *non-traditional background*. These professionals may come from diverse fields with less focus on quantitative tasks, such as psychology, music, or even journalism. This unconventional background can provide them with unique perspectives and creative problem-solving abilities, enriching the field of data science with their varied experiences.

Off-the-beaten path-ers possess a wide range of educational and professional backgrounds, which equip them with diverse skills and knowledge. They may have initially pursued a career in a different domain before discovering their passion for data science. This varied experience often results in a broader, interdisciplinary approach to problem-solving, allowing them to draw connections and insights that might be overlooked by their more traditionally trained peers. For example, off-the-beaten path-ers might approach the problems within ML and **artificial intelligence** (**AI**) ethics (a topic of increasing relevance within AI) differently than the traditionalist or domain expert. They may also regard ML and AI as tools to create a better world by tackling humanitarian issues such as disaster response, public health, food security, and human rights. Furthermore, AI may also be of interest to civil engineers with an interest in smart cities or political science majors with detecting implicit biases in the criminal justice system.

With their unconventional backgrounds, off-the-beaten path-ers bring a unique perspective to data science, enabling them to tackle problems from a different angle. Their creativity and innovative thinking can lead to the development of new methods, models, or visualizations that challenge the status quo and push the boundaries of what is possible in data science. This outside-the-box thinking is valuable, especially when addressing complex or novel challenges.

Also, with their unique backgrounds, off-the-beaten path-ers are well equipped to collaborate with professionals from various disciplines, leveraging their distinct perspectives to solve complex problems. Their ability to work effectively with interdisciplinary teams can lead to the development of innovative solutions that combine the strengths of multiple fields, driving growth and success for the organization. To facilitate working with different backgrounds, they often have to communicate complex ideas and insights effectively to diverse audiences. Off-the-beaten path-ers often understand the importance of storytelling in data science, using data visualizations and narratives to convey their findings clearly and compellingly. This skill enables them to bridge the gap between technical experts and non-technical stakeholders, facilitating collaboration.

In conclusion, if you have come to data science as an off-the-beaten path-er, we recommend positioning yourself in job interviews as someone who is adaptive and can bring your unique perspective to facilitate creative problem-solving. Additionally, highlight any abilities to communicate and collaborate.

As the field of data science continues to expand, the diversity of its professionals will only increase. The traditionalist, domain expert, and off-the-beaten path-er each bring unique strengths and perspectives. Of course, these are just generic groupings of data science professionals and you may be a mix of all of these profiles. Embracing your individual strengths will allow you to best position yourself in a data science interview.

Nonetheless, while all of these paths have their benefits, none of them are without barriers. A common misconception in data science is there is a *perfect* path, or one that's comprehensive such that the path with be without bottlenecks. While it is true that some paths have advantages over others, they each have gaps to address. While some of these gaps are flavor- or path-specific, they all share one: getting the first data science job.

Tackling the experience bottleneck

So, you want to be a data scientist? Welcome to *The Hunger Games: Data Science Edition*!

While that may sound like an exaggeration, the increasing demand for data scientists has turned the interview process into a battleground for candidates with various backgrounds and expertise.

But fear not – just as with *The Hunger Games*, the odds can be in your favor.

The fact that there is competition should not scare you away from entering the field. You've already shown your interest and commitment by reading this book, and as you progress through it, you'll learn how to prepare for data science interviews, regardless of your background. In addition, we will share strategies to fill gaps in your experience to make you a stronger candidate. Remember – you have your own set of strengths and weaknesses. You can come out on top by focusing on your gaps and understanding your unique skills.

Believe it or not, it's incredibly common for candidates to have gaps in their experience. In the next couple of sections, we will review two familiar sources of experience gaps: academic and work experience gaps. In addition to noting these gap areas, we will give you suggestions on how to close them.

Academic experience

One common gap in a job candidate's experience is their academic background. Employers may favor candidates with formal degrees in data science, computer science, or a related field, making it challenging for those without a traditional academic background to stand out. You may not be an engineer or a programmer by trade, but you understand math or computers but have yet to get into the details of hypothesis testing. There's no need to worry. The first step in addressing gaps in your academic background is identifying them. Reflect on your education and experience, and ask yourself the following questions:

- In which areas of data science do I feel the least confident?

- To which technologies or concepts do I need more exposure?

- Which topics or tasks do I struggle with the most during interviews or when working on projects?

- What models are commonly needed for the job that I want?

Once you've identified your gaps, you can create an action plan to address them effectively. Here are several methods to help you fill the academic experience gap and strengthen your data science candidacy:

- *Pursue relevant certifications*: Obtain certifications in data science, ML, AI, or related fields from reputable organizations or platforms (for example, DataCamp, Codeacademy, Sololearn, Alison, Udemy, Udacity, Google certifications, and so on). These certifications can help you gain credibility, showcase your expertise, and demonstrate your commitment to learning.

- *Attend workshops and boot camps*: Participate in workshops, boot camps, or short-term courses that provide hands-on experience in data science techniques and tools. For example, Meetup.com and LinkedIn are useful sites for identifying local or virtual data science groups. This will not only help you enhance your skills but also allow you to connect with other professionals in the field.

- *Leverage Massive Open Online Courses (MOOCs)*: Enroll in MOOCs from top universities or platforms to learn data science concepts and techniques. Common websites include *Coursera* and *edX*. These courses can help you build a strong foundation in the subject and supplement your non-traditional academic background.

- *Build a strong portfolio*: Create a robust portfolio that showcases your data science projects, coding skills, and problem-solving abilities. Highlight your unique perspective and how your non-traditional background has contributed to your approach to data science.

- *Network with data science professionals*: Connect with professionals in the data science field through networking events, online forums, or social media platforms such as LinkedIn. This can help you gain insights into the industry, learn about job opportunities, and build relationships that can lead to mentorship or job referrals.

Resources, such as books, online courses, and tutorials, help you gain the necessary knowledge. Develop a realistic timeline for completing any of these activities and don't become overwhelmed by the vast availability of online courses. Setting achievable goals and being patient with yourself is important when developing your learning plan. Remember – data science is a vast field, and it takes time to become proficient. Set a dedicated time to work on your learning plan. In addition, engage with the data science community through forums, social media, and networking events to learn from others and stay motivated.

Work experience

Another common experience gap for candidates is related to work experience. Entering the data science field can be challenging, particularly when faced with the work experience bottleneck. Employers often seek candidates with prior experience, creating a catch-22 for aspiring data scientists: you need experience to get a job, but you need a job to gain experience! This section will explore common reasons for gaps in a work background and provide strategies to help you overcome the work experience bottleneck.

There are several reasons why your work background might not perfectly align with what an employer is looking for, such as a career transition from a different field; you may be a recent graduate with limited or no full-time experience, or you may have employment gaps due to personal reasons (for example, caregiving, health, travel) or have done freelance or contract work, which may not be perceived as consistent or relevant experience.

Understanding the reasons behind work background gaps is essential for crafting a compelling narrative and demonstrating your value to potential employers. Here are several methods to help you fill the work experience gap and strengthen your data science candidacy:

- *Personal projects*: Develop and showcase personal projects demonstrating your skills, creativity, and problem-solving abilities. Choose projects that align with your career interests or target industries. This will help build your portfolio and show your passion and commitment to the field.

- *Internships, co-ops, fellowships, and apprenticeships*: Seek internships, co-ops, or apprenticeships to gain hands-on experience and make valuable connections in the industry. These opportunities can provide a foot in the door, allowing you to learn from experienced professionals and build a network that can lead to future job prospects. There are even some online internships. For example, *Forage* offers virtual experiences hosted by top companies including JPMorgan Chase, Walmart, KPMG, Lyft, Red Bull, PWC, Accenture, Deloitte, GE, and more. Many tech companies such as Microsoft, Amazon, and Google offer many apprenticeships for recent graduates and professionals. Some organizations offer online fellowships, such as Correlation One and Insight Fellows.

- *Freelance and consulting work*: Offer freelance or consulting services to businesses and organizations, even if on a *pro bono* basis. This allows you to gain practical experience, enhance your skills, and build a track record of success. In addition, it demonstrates your ability to work with clients and solve real-world problems. Websites include *Upwork, Fiverr, FlexJobs*, and so on.

- *Online competitions and hackathons*: Participate in data science competitions and hackathons, such as those hosted on *Kaggle* or *DrivenData*. These events allow you to work on challenging problems, collaborate with others, and showcase your skills to potential employers.

- *Open source contributions*: Contribute to open source projects related to data science, ML, or AI. This improves your technical skills and demonstrates your ability to collaborate with others and contribute to the broader data science community.

By employing these strategies, you can overcome the work experience bottleneck and position yourself as a strong candidate in the data science job market. Remember – persistence and adaptability are key to success. Stay focused on your goals, seize opportunities to learn and grow, and, ultimately, you'll break through the work experience barrier to land your dream data science job.

Now that you've had a proper introduction to bottlenecks that you might encounter, as well as methods and resources to address them, let's gain a better understanding of the skills and competencies that are expected of you. After reviewing both hard skills and underrated soft skills, you will be able to isolate your competency gaps, which will not only help you identify which resources to leverage but will also help you navigate this book in a more pointed and goal-oriented fashion. While it is encouraged to review the book in its entirety, you can prepare for sections that might require more attention.

Understanding expected skills and competencies

Here's the deal – the interview is a critical component of the data science job application process, where you can showcase your skills, knowledge, and personality to potential employers. The interview process is crucial for several reasons:

- Employers can assess your technical skills, problem-solving abilities, and critical thinking

- It lets you demonstrate your communication skills, teamwork, and cultural fit

- It allows you to ask questions and gather information about the company and role to ensure it aligns with your career goals and values

- Preparing for the interview is essential to stand out in the competitive job market and secure your dream role

Preparing for the data science interview is essential to success. In fact, it's one of the most useful activities that you can do for your career. This is not only true for prospective data scientists looking to land their first job in the field but also for well-seasoned data scientists who wish to stay on top of new techniques and technologies. In later sections of this book, we will help you prepare by reviewing the most common data science interview topics, including technical and case study questions. In addition, we will give you problems to practice your problem-solving skills, coding, and data manipulation techniques. Including these activities, you should also prepare by researching the company, its culture, products, and industry trends. Additionally, prepare questions to ask the interviewer to demonstrate your interest and engagement.

For now, know that most data science interviews consist of two primary areas: technical (hard) skills and non-technical (soft) skills. Each area serves a different purpose and requires distinct preparation strategies. The technical portion assesses your knowledge and skills in data science, programming, statistics, and ML. For example, it may include coding exercises or algorithmic questions, data manipulation and cleaning tasks, statistical analysis or hypothesis testing questions, and ML model selection and evaluation problems. Meanwhile, the non-technical portion evaluates your communication skills, problem-solving skills, and ability to work in a team. It may involve questions about your past experiences and accomplishments, situational or problem-solving scenarios, discussion of your strengths, weaknesses, and work style, and exploration of your motivations and career aspirations.

Mastering the data science interview is a crucial skill that can make or break your career. While we don't win them all, studying for these interviews can feel like preparing for a marathon. This is especially true when you have to prepare for multiple interviews and/or take-home assignments. The key to breaking into the data science field is building strong foundations in expected skills and competencies. By excelling in the interview process, you can leave a lasting impression on potential employers and increase your chances of receiving a job offer. Furthermore, understanding the interview's structure thoroughly prepares you for both technical and non-technical portions, and by effectively highlighting your strengths and skills, you'll be well on your way to success in the data science field.

Let's take a deeper look into what's included in the hard and soft skills expected of a prospective data scientist. After the review, you will have a clearer concept of the proficiencies you will learn throughout this book.

Hard (technical) skills

To excel in a data science role, you must possess a strong foundation in various hard technical skills. These skills enable you to effectively manipulate, analyze, and interpret data and develop and deploy ML models. In this section, we'll discuss the essential hard technical skills required to be successful in a data science position:

- *Programming languages*: Proficiency in programming languages is crucial for data manipulation, analysis, and visualization. The most popular languages in data science are:

  - *Python*: A versatile, high-level programming language with extensive libraries and tools for data science, such as NumPy, Pandas, Matplotlib, and scikit-learn (we cover some key Python skills later in the book).

  - *R*: A language specifically designed for statistical computing and graphics, offering a wide range of packages for data manipulation, visualization, and modeling.

- *Data manipulation and cleaning*: Data scientists often work with raw, messy, or incomplete data. Therefore, you must be skilled in data preprocessing, cleaning, transforming, and organizing data to prepare it for analysis or modeling. Proficiency in SQL is often needed to pull data from databases and clean and prepare it.

- *Data visualization*: Data visualization represents data in a graphical format to effectively communicate insights and trends. Essential data visualization skills include creating clear and informative visualizations using tools such as Matplotlib, ggplot2, or Tableau and selecting appropriate visualization types based on the data and the intended audience. Effectively communicate insights and findings through visual storytelling.

- *Statistics*: A strong foundation in statistics is vital for making data-driven decisions and interpreting results. Key statistical concepts and techniques in data science include descriptive statistics, which summarize and describe data using measures such as mean, median, mode, variance, and standard deviation. Additionally, a candidate must know inferential statistics, which draws conclusions about populations or relationships based on sample data using techniques such as hypothesis testing and confidence intervals. Also, probability theory is about understanding the likelihood of events and their relationships, including concepts such as conditional probability, independence, and Bayes' theorem.

- *ML*: ML involves training algorithms to learn from data and make predictions or decisions. Essential ML skills include:

  - *Supervised learning (SL)*: Building models to predict target variables based on input features. Some SL techniques that you should understand before your data science interview include linear regression, logistic regression, and decision trees.

  - *Unsupervised learning (UL)*: Discovering patterns or structures in data without labeled targets. Techniques such as clustering, dimensionality reduction, and anomaly detection are important to understand before your data science interview.

  - *Model evaluation*: Assessing model performance using metrics such as accuracy, precision, recall, F1 score, and **area under the curve (AUC)**.

- *Cloud computing platforms*: Services such as AWS, Azure, or Google Cloud provide scalable resources for data storage, processing, and ML. More and more organizations are adopting these platforms, and they will likely require you to know how to perform data science activities using them, although most services offer certificates to show your proficiency in using their service.

It's essential to continuously refine and update your skills to stay competitive in the rapidly evolving field of data science. Engage in ongoing learning, attend workshops, and participate in online courses or boot camps to keep your technical skills sharp and relevant.

Soft (communication) skills

While hard technical skills form the foundation of a data scientist's expertise, soft skills are equally important in ensuring success in the role. Soft skills are non-technical, interpersonal abilities that help you navigate professional relationships, collaborate with team members, and effectively communicate your insights. This section will discuss essential soft skills required to excel in a data science position:

- *Curiosity and continuous learning*: A successful data scientist must possess a curious mindset and commitment to ongoing learning. Fostering curiosity and continuous learning includes staying informed about industry trends, new tools, and techniques. Additionally, seek feedback from peers, mentors, and supervisors to identify areas for growth. Finally, engage in professional development activities, such as attending conferences, workshops, or online courses.

- *Communication*: Effective communication is critical for data scientists, as it enables you to explain complex concepts and insights clearly and concisely, tailored to your audience. Also, it is crucial that you present your findings and recommendations to both technical and non-technical stakeholders.

- *Teamwork and collaboration*: Data scientists often work in multidisciplinary teams, collaborating with engineers, analysts, product managers, and other stakeholders. Essential teamwork and collaboration skills include active listening and consuming others' perspectives, needs, and ideas. Adaptability is also essential for collaboration, adjusting your approach and priorities to accommodate changes in team dynamics, project requirements, or goals.

- *Problem-solving*: Data scientists must tackle complex, real-world problems by breaking them into smaller components, analyzing available data, and developing appropriate solutions. Key problem-solving skills include analytical thinking, where you identify patterns, trends, and relationships in data and understand the underlying structure of problems.

- *Time management and organization*: Effective time management and organization skills are crucial for managing multiple tasks, meeting deadlines, and prioritizing work. To excel in these areas, consider setting clear goals and objectives for both short-term and long-term projects. Also, create a structured schedule that gives time for different tasks and priorities. Finally, you should regularly assess progress, adjust plans as needed, and learn from past experiences.

These hard and soft skills are what make up a comprehensive data scientist who is not only equipped to use mathematical and computational techniques to tackle business questions but is also skilled in effectively managing multiple projects, deliverables, stakeholder expectations, and tight deadlines. While data scientists are typically not the most client-facing role in an organization, the best data scientists stand out when they have strong interpersonal skills to collaborate and communicate questions, requirements, caveats, how models work, and how to interpret results. After all, your work is only as good as how it's communicated.

Exploring the evolution of data science

The field of data science continues to evolve, both in terms of the tools used and the type of work conducted. This evolution is driven by advancements in technology, the increasing availability of data, and the growing demand for data-driven insights in a wide range of industries. As a result, it is critical for those interested in entering the field to not only learn fundamental techniques and technologies of data science but also to stay diligent and current on new developments and technologies.

New models

One of the most significant ways in which the field of data science is evolving is through the development of new ML and AI algorithms and techniques. As AI continues to become more sophisticated, data scientists are able to build more accurate and powerful predictive models that can be used to solve a wide range of complex problems. This includes the implementation of methods borrowed by other fields in industry and academia such as process improvement, operations research, game theory, network/graph analytics, and deep learning techniques.

It goes without saying, but developments such as the **large language models** (**LLMs**) used in ChatGPT are expected to have a profound impact on how data scientists work. For example, LLMs in **integrated development environments** (**IDEs**) have the potential to expedite the writing of code. This is comparable to the development of **open source software** (**OSS**) packages, which have already increased productivity for programmers.

New environments

Another way in which the field of data science is evolving is through the increasing use of cloud-based technologies and platforms. Virtualization and serverless technologies have provided data scientists with the ability to access powerful computing resources and scalable data storage, making it easier and more cost-effective to work with large datasets. Thus, cloud computing has revolutionized the data science landscape by offering unprecedented opportunities and transforming the way organizations approach data analysis and ML. With these advancements, data scientists have overcome traditionally inconvenient constraints such as hardware limitations, scalability challenges, and resource allocation. Now, data scientists can create multiple **virtual machines** (**VMs**) on a single physical server, enabling efficient utilization of computing resources.

For example, serverless technology simplifies model deployment and management of software applications, as it eliminates the need for infrastructure provisioning and automatically scales resources based on demand. Cloud computing platforms such as AWS, Microsoft Azure, and GCP have dominated the **Infrastructure-as-a-Service (IaaS)**, **Platform-as-a-Service (PaaS)**, and **Software-as-a-Service (SaaS)** spaces and democratized access to high-performance computing, storage, and specialized tools, empowering data scientists with immense computational power at their fingertips. They offer powerful frameworks such as Google Cloud AI Platform and Azure Machine Learning, which can train complex models on massive datasets without investing in expensive hardware. Additionally, cloud-based data lakes such as AWS Simple Storage Service (S3) or Azure Data Lake Storage (ADLS) provide scalable and cost-effective storage solutions for large-scale data processing and analysis.

Overall, virtualization, serverless technology, and cloud computing have dramatically expanded the capabilities and reach of data science, enabling more efficient and scalable data analysis, fostering innovation, and accelerating the development of AI-driven solutions across industries.

New computing

Improvements in computational power will also continue to drive the field forward. As datasets continue to grow in size and complexity, and as AI algorithms become more sophisticated, data scientists require more powerful computing resources to process and analyze data. This has led to the development of specialized hardware and software tools designed specifically for data science, such as GPUs, and distributed computing frameworks such as Hadoop and Spark. In addition, many data scientists are now turning to cloud-based computing platforms such as AWS and Google Cloud to access scalable computing resources on demand.

With the rapid pace of technological advancement in the field of data science, it is critical for data scientists to stay up to date on the latest developments in computational power and to have the skills and knowledge necessary to take advantage of these resources.

New applications

In addition to these technical advancements, the field of data science is also evolving in terms of the industries and applications where it is being used. Data science is now applied to a wide range of fields, from healthcare and finance to transportation and logistics. As a result, data scientists must adapt to new industries and domains and be able to apply their skills and techniques to solve new and unique problems.

Given the rapid pace of change in the field of data science, it is critical for individuals interested in entering the field to stay diligent and current on new developments and technologies. This requires a commitment to ongoing learning and professional development, as well as an openness to new ideas and approaches. By staying up to date on the latest advancements in the field, data scientists can ensure that they remain competitive and are able to deliver value to their organizations and clients.

Summary

In this chapter, you've learned about the modern data science landscape, what the role entails, what skills and competencies are expected of a prospective candidate, and the most common paths to becoming a data scientist. Furthermore, you've learned about the multi-faceted functionality of data science and how it leads to a diverse workforce of data scientists with different specialties and backgrounds.

With this in mind, you may determine what your path might look like or what knowledge gaps you hope to close. Whichever the case, you are now prepared to move forward with your interview preparation.

In this next chapter, we will begin the data science job search journey by mentally (and emotionally!) prepping you for the road ahead. We'll discuss some underrated tips on how to identify the right job opportunity, where to find it, how to create a stand-out application, and how to stay ahead of the curve in a sea of evolving technology, project portfolios, and resumes. We hope that you are as excited as we are to get started!

References

- [1] *Data science* from *Wikipedia*: https://en.wikipedia.org/wiki/Data_science

- [2] *Is Data Scientist Still the Sexiest Job of the 21st Century?* by *Thomas H. Davenport* and *DJ Patil*, from *Harvard Business Review*: https://hbr.org/2022/07/is-data-scientist-still-the-sexiest-job-of-the-21st-century

- [3] *Data Scientists* from *U.S. Bureau of Labor Statistics*: https://www.bls.gov/ooh/math/data-scientists.htm#tab-1

- [4] *The Digitization of the World* by *David Reinsel*, *John Gantz*, and *John Rydning*, from *International Data Corporation*: https://www.seagate.com/files/www-content/our-story/trends/files/idc-seagate-dataage-whitepaper.pdf

2

Finding a Job in Data Science

Now that you have decided to grow your career in data science, let's go and get one!

This chapter will cover effective job search strategies, including how to mentally prepare yourself and develop an effective resume and work portfolio. Our goal is to position you to be successful in your job search. In addition, we provide tips from insiders.

After completing this chapter, you will know how to properly develop a data science job search plan and strategy, complete with a stellar resume and cover letter to attract potential employers and an impressive project portfolio. You will also have a grasp on when and how to source jobs through networking and online job posts, and how to stay ahead of the curve with new technology skills.

In this chapter, we will cover the following topics:

- Searching for your first data science job
- Constructing the golden resume
- Prepping for landing the interview

Searching for your first data science job

Embarking on your data science job search requires careful preparation, diligence, patience, and thick skin. Mental readiness is as vital as technical expertise, with the search often becoming a marathon rather than a sprint. Therefore, maintaining composure and perseverance in this demanding field is paramount for success. To achieve this mindset, it is pertinent that you leverage effective job-hunting skills.

An effective job hunt leverages several tools and resources at your disposal. Job board sites are crucial, functioning as bridges between you and potential employers. Learning how to navigate these platforms effectively can convert them from daunting job databases into your personal gold mines. Equally crucial is a professionally curated portfolio that showcases your technical acumen, problem-solving prowess, creativity, and passion for data science. That's right – landing a data science job will require more than technical prowess. It's a bit of a science and art form, requiring some creative and clever strategies.

The art of applying for jobs combines all of your job preparation efforts – resume writing, strategic networking, strategic applying, and so much more. The process goes beyond merely clicking a button or sending an email – it demands strategic timing, tailoring your applications, and aligning your unique skillset with the vision and goals of prospective companies.

This chapter aims to provide a comprehensive guide to successfully navigate these steps, equipping you with the knowledge, strategies, and tips necessary for an effective job search in the data science field. This begins with an exploration of the mental journey ahead.

Preparing for the road ahead

Embarking on a job search can stir up a whirlwind of emotions. However, the initial excitement of seeking a new role in data science can quickly be tempered by the reality of the challenges ahead. Luckily for you, this job search process often follows a predictable emotional cycle (as data scientists, we love predictability!). Much like forecasting, it allows us to take a peek at the future and plan accordingly.

The journey typically begins with a sense of optimism and excitement at the prospect of new opportunities. Kourtney Whitehead, a career expert and the author of *Working Whole*, says, "*Don't try to temper your expectations or assume your positive attitude is naïve. In fact, the hopefulness you feel in the early stages of a job search is in recognition of the true opportunity that stands before you*"[1].

However, as time progresses and the realities of a competitive job market set in, feelings of frustration, disappointment, and self-doubt can surface. There may be periods when your applications seem to disappear into a void, or you might face rejection after investing significant time and energy into interviews. These experiences can feel disheartening and may cause emotional lows, but embracing the emotional cycle as a normal part of the job search process is the first step in preparing mentally for the journey ahead.

Strategies for emotional resilience

Here are some strategies to help you maintain emotional resilience during your job search:

- *Identify your motivation*: Understanding why you want to become a data scientist will help you focus on the end goal and motivate you during difficult times.

- *Maintain perspective*: Remember that your worth is not defined by your job or the number of rejections you receive. The job search is just one aspect of your life, and rejection is a common part of the process that even the most successful professionals have faced.

- *Self-care*: Prioritize activities that help you relax and de-stress. This could be exercise, meditation, spending time with loved ones, or pursuing a hobby. These activities can help you maintain balance and prevent burnout.

- *Support network*: Surround yourself with supportive friends, family, or mentors who understand your journey and can offer encouragement during low points.

- *Celebrate small wins*: Received a callback? Made it through a tough coding challenge? Celebrate these victories. They indicate progress and can boost your confidence.

- *Reflect and learn*: Use rejections as opportunities for growth. Request feedback where possible, reflect on your performance, and identify areas for improvement.

Staying patient and persistent

Patience and persistence are vital in navigating the ups and downs of a job search. Here are some strategies to cultivate these traits:

- *Set realistic expectations*: Remember that landing a job, especially in a competitive field such as data science, can take time. Prepare for the possibility that your job search may be a marathon, not a sprint.

- *Consistent effort*: Decide how much time you want to dedicate in a week or each day to your job search activities, such as networking, applying for jobs, and improving your skills. Set a dedicated time for these activities. Consistency can help you maintain momentum and progress.

- *Flexible approach*: If you're not getting the desired results, be willing to adjust your strategy. This could mean broadening your job search, improving your resume, or learning a new skill.

- *Stay informed*: Keep up to date with the latest trends and demands in the data science market. This can help you identify new opportunities and stay motivated. We will discuss this more later in this chapter.

- *Fight off procrastination*: Thinking about a new job is easier than working on your resume or online profile. Use the motivation you identified earlier to help get you started and avoid putting it off for later. Remember, you can't get your next data scientist role until you start.

How to get started when life is so busy

Searching for a job sometimes feels like a full-time job in itself. It's especially daunting if you currently have a job; repetitively completing applications by reentering the same information into a portal and preparing for the actual interview can take hours. As a result, you may begin to question your resolve in finding your new position. But stick with it! You have already shown your commitment by reading this book; you must continue even through the challenging moments.

This is where maintaining a consistent effort during your job search is critical.

Firstly, it helps keep you engaged and prevents inertia from settling in. Job hunting can often feel like a numbers game, but your odds of success increase with every application you submit, every networking event you attend, and every new skill you learn. Designating a specific time each day or week for job search activities can create a routine that makes the process feel less overwhelming and more manageable.

Secondly, consistency demonstrates a key professional attribute – resilience. It is the ability to stay focused and committed in the face of challenges, a trait highly valued in data science, where problems can be complex and solutions may not be immediately apparent.

Lastly, consistent effort allows you to stay current with the dynamic nature of the job market. By regularly checking job boards, networking, and improving your skills, you stay in tune with the evolving demands and trends in the data science industry. This continual engagement and adaptability can give you a competitive edge in the job market.

Therefore, preparing emotionally for your job search is as important as updating your resume or brushing up on your technical skills. You can navigate your job search journey with a healthier and more balanced mindset by acknowledging the emotional highs and lows, practicing emotional resilience, and cultivating patience and persistence. Remember, every step you take brings you closer to your goal, and every challenge you overcome makes you a stronger candidate.

Finding job boards

After setting yourself up for success by preparing mentally for the journey ahead, one of your next steps is to begin the job search. This is commonly accomplished by tapping into personal and professional networks and scanning job boards.

Job board sites such as LinkedIn, Glassdoor, and Indeed have revolutionized the job search process, providing a wealth of opportunities and resources at your fingertips. These platforms are not just avenues for applying for jobs but also powerful tools for research, networking, and gaining insights into the data science industry. This section will guide you on effectively utilizing these platforms beyond the simple **Apply** button.

Sampling job board sites

Each job board site offers unique features to aid your job search. Here are just a few:

- *LinkedIn*: Primarily a professional networking social media site, LinkedIn hosts a robust job board that allows you to connect with professionals in the industry, join relevant groups, follow companies of interest, and showcase your professional brand.

- *Glassdoor*: Glassdoor is known for its company reviews, salary reports, and interview insights provided by employees. It also features a job board that can be filtered by location, job title, and company.

- *Indeed*: Indeed is a comprehensive job board aggregating job postings from various websites. It also allows users to upload resumes and apply for jobs directly.

- *Handshake*: Handshake is a job site specifically tailored toward college students. It lists job and internship opportunities and provides opportunities to meet with company recruiters.

- *Built-In:* Built-In is a job board known for its tech startup postings, which make great options for professionals looking for smaller and/or newer companies.

- *Dice:* Dice is a job board that specifically posts jobs within the tech sector.

- *Fishbowl:* Fishbowl isn't a job board, but rather a social media app where professionals can engage in career discussions, anonymously. This format encourages honest and transparent discussions on company culture, compensation, and many other topics. It's a great resource if you want the "insider's" perspective on working for a company.

Each of these sites has its own uniqueness, but they all offer the ability to alert you as new roles are added to their site that match your profile. Make sure that you utilize this ability, as it is a great way to continue to have leads come in. Depending on how aggressive you are with your search, you can set these alerts to occur anywhere between once a month to once a day. In either case, be sure to utilize relevant keywords and job search criteria such as **Location**, **Format** (remote versus hybrid versus on-site), **Type** (full-time versus part-time versus contract), **Years of Experience**, and others.

Using job board sites for research

Job board sites can be a gold mine of information to inform your job search strategy. Here's how:

- *Understand the market*: Regularly browsing these sites can provide insights into the types of roles available, the most in-demand skills, and the companies hiring data scientists.

- *Analyze job descriptions*: Studying job descriptions can help you understand the qualifications, skills, and experience employers seek. This can guide your learning pathway and help tailor your applications. We will talk more about this later in the chapter.

- *Company research*: Company pages on LinkedIn, reviews on Glassdoor, and employee discourse on Fishbowl can give you a sense of the company culture, values, and work environment. This can help you identify organizations aligned with your career goals and values.

Other job site tips

Here are some more tips for using job sites:

- Clearly define your job search criteria. Determine the industry, location, and other specific requirements that align with your career goals.

- Consider applying for data science jobs that more closely align with your academic and/or professional background. For instance, if you studied geology, consider finding geospatial or environmental data science roles. Similarly, if you have experience in the healthcare industry, consider finding data science roles in pharma, insurance, or informatics.

- Keep your profile up to date with the latest information, such as work experience, education, and skills. This increases your chances of being contacted by recruiters.

- Make use of advanced search filters available on job search websites to refine your search based on factors such as location, salary, experience level, and job type. Try to nail down a few keywords that are more likely to come up in the roles that you seek.

- Before applying for a job, research the company to gain insights into its culture, values, and reputation. This information will help you tailor your application and prepare for interviews.

- Read job descriptions carefully and follow the application instructions provided by the employer. Missing out on specific requirements can lead to your application being overlooked.

- If the website allows you to upload a profile picture, use a professional-looking image that presents you in a positive and suitable manner.

- Research salary ranges. Use job search websites to research salary ranges for positions in your desired industry and location. This information can help you negotiate a fair compensation package during the hiring process.

- On LinkedIn, do not fall for the habit of only applying to **Easy Apply** roles. The easier it is to apply, the more competition you will have. For this reason, use job board sites such as ZipRecruiter very sparingly.

- Focus most of your time applying to roles that have been live for no longer than one week. Unless recruiters are behind schedule, most of them have gathered a healthy database of applicants to interview. Prioritize roles by how new they are. As the saying goes, "The early bird gets the worm."

Networking and building connections

Networking plays a pivotal role in the tech and data science industry, serving as a crucial gateway to professional growth, collaboration, and opportunities. In this dynamic and rapidly evolving field, building a robust network allows individuals to connect with like-minded professionals, experts, and mentors who can provide valuable insights and guidance. Through networking, professionals can expand their knowledge base, stay updated on the latest industry trends, discover new tools and technologies, and make meaningful connections with other professionals.

Moreover, networking facilitates the exchange of ideas, fostering innovation and creativity. It opens doors to potential job opportunities, partnerships, and collaborations, enabling individuals to advance their careers and make meaningful contributions to the industry.

In the tech and data science industry, where staying ahead of the curve is essential, networking acts as a catalyst for professional success, providing a platform for continuous learning, support, and growth.

Here are some ways to harness this potential:

- *Connect with professionals*: Don't hesitate to connect with other data scientists. A personalized connection request outlining your interest in their work or the field can go a long way. Reach out to recruiters or HR professionals directly to express your interest and inquire about potential opportunities. Build meaningful connections and seek referrals whenever possible. The more you can streamline this process while creating and maintaining meaningful relationships, the better. Google Sheets now offers a ChatGPT plug-in, which allows you to pen personalized introductory emails based on each row of professional career information. Zapier can also be used to perform a similar task.

- *Informational interviews*: Reach out to connections for informational interviews. This is a non-threatening way to learn about their role and journey and gain valuable advice. Remember, this is not an opportunity to ask for a job but to learn and build a relationship. Although this will make the interviewee aware of your background, you shouldn't hesitate to share that you are seeking a job.

- *Engage with content*: Commenting on posts, sharing articles, and contributing to discussions can increase your visibility and present you as an engaged member of the data science community.

- *Join groups*: LinkedIn groups can be a source of industry news, discussions, and job postings. Participate actively to gain and share insights. Other sites and apps such as Slack, Discord, and Meetup allow you to meet professionals with similar interests by joining topic-based groups. These communities often share job-hunting tips, job postings, recruitment processes, and networking opportunities that can lead to referrals. Sites such as Blind and Fishbowl have groups based on interests and company, allowing users to engage anonymously, which encourages users to share information that they otherwise may not. Additionally, apps such as Slack and Discord also allow you to join topic-based communities for networking opportunities.

- *LinkedIn Premium*: At the time of this writing, LinkedIn offers some paid features for job seekers. For jobs applied for via their website, this includes the ability to view how many other candidates have applied for the job and some of their skills, allowing you to understand your competition. Additionally, you have a chance to see who is viewing your profile; this information will enable you to network with recruiters who come across your profile.

Finding job leads

As previously mentioned, your connections and the company pages that you follow could generate job leads. While many companies' job postings can be located on job board sites, many more jobs are never posted. According to Flex Jobs, roughly 70–80% of job openings never hit the internet [2].

However, connections who know that you're job hunting might share relevant internal opportunities at their companies. This is where your personal and professional networks really make a difference. Letting people know you are looking for a job goes way beyond enabling the **Open for Work** banner on your LinkedIn profile.

Although networking may feel weird at first, there are countless books out there that teach you how to network effectively and naturally. In some cases, you may make lifelong connections and acquaintances, which is an enriching experience all on its own. In either case, networking increases your chances of finding a data science job with fewer applicants, given that many job openings never go public.

COVID-19 leads to more remote work

The COVID-19 pandemic has impacted our society in many ways, including where and how work is performed. Although remote work wasn't new at the time, the start of the pandemic forced many companies to adopt remote and hybrid work formats for their employees. As a result, many of those companies found that they could run their organization successfully this way. The tech industry, as well as technical workers, were perhaps those who benefited the most from having this option.

As hiring managers are now more comfortable with individuals working remotely, there has been an increase in remote work positions. Remote work does not suit everyone. However, candidates who are either uninterested or unable to relocate for a new job can pursue more roles through remote work. The pool of available job leads has increased, and many job board sites allow you to specifically filter to remote and hybrid jobs. In many cases, applications are beginning to ask for the candidate's preference. However, it is worth noting that the increase in remote work has not outpaced the number of employees interested in landing these roles. With that said, be sure to apply for on-site and hybrid roles, as remote roles are more competitive.

Job board sites are powerful allies in your data science job search. By leveraging these platforms for research, networking, and job leads, you can make informed decisions and uncover previously inaccessible opportunities through traditional job search methods. Remember, in the digital age, your online presence and activities speak as loudly as your resume.

Interpreting job descriptions

Navigating the job market can often feel daunting, particularly when you come across a job description that lists an array of qualifications, some of which you might not possess. So, here's a crucial piece of advice: remain confident, even if you don't meet 100% of the job description's requirements. It's perfectly acceptable, and indeed common, to apply for roles even when you don't meet every single criterion. In fact, it's advised to apply for jobs even if you only meet 70–80% of the job requirements. Furthermore, some argue that if you meet 100% of a job description, it leaves little room for growth, to which many recruiters attribute high attrition [3].

In reality, job descriptions are typically an employer's wish list, outlining the ideal candidate's skills and qualifications. In most cases, this results in job descriptions that list more programming languages and technical frameworks than you actually need to know. Thus, recruiters recognize that finding a candidate who checks all the boxes is quite rare – potentially non-existent. Employees often look for high potential and a willingness to learn. If you can show that you are eager to grow, adapt, and have a solid foundation on which you can build the required skills, many employers will consider your application.

Being passionate about the job role can sometimes compensate for some lack of experience or skills. If you convey your enthusiasm effectively in your application and during the interview, hiring managers are likely to consider you seriously for the role. They understand that a passionate candidate is likely to be motivated, dedicated, and willing to learn – qualities that can sometimes outweigh specific technical skills.

Remember, the worst-case scenario is that you're not selected for an interview, but if you get the opportunity to interview, it's your chance to explain why you're a great fit for the role, regardless of not meeting all of the requirements. You can highlight transferable skills from your previous experiences, showcase your learning ability, and express your passion for the role and the industry.

Don't let a job description deter you from applying for a role that you're genuinely interested in. Believe in your potential and give yourself the chance to share your candidacy. After all, the job search journey is not just about the destination but also about the valuable lessons and skills you gain along the way.

Beginning to build a standout portfolio

Oftentimes, technical roles require an extra layer to an applicant's candidacy. As you begin to search for jobs, you might notice that many of them request a link to a portfolio.

A **portfolio** of data science projects is a repository that highlights a spectrum of your technical aptitude and potential. Portfolios can differentiate you from other candidates during your job search. A well-crafted portfolio also showcases your creativity, problem-solving abilities, learning journey, and passion for the field. For junior and entry-level data scientists with limited professional experience, portfolios are highly recommended.

This section will provide some tips and pointers on how to build a compelling data science portfolio that can give you a competitive edge in the job market.

Starting your portfolio

If you're new to data science, you may wonder what to include in your portfolio or where to host it. Here are a few options:

- *Coursework projects*: If you've completed a data science degree or boot camp, you likely have a collection of projects you've worked on. Choose those that best demonstrate your skills and make sure they're polished and well documented.

- *Personal projects*: Working on a project related to a topic you're passionate about can make the process enjoyable and result in a unique portfolio piece. This could be analyzing sports statistics, election data, or financial trends. By using public datasets, you can showcase your ability to extract insights from data. Just try to avoid overly used datasets often hosted on Kaggle.

- *New techniques or algorithms*: Whenever you learn a new technique or algorithm, consider creating a small project to apply what you've learned. This demonstrates your commitment to continual learning while solidifying your new knowledge. Over time, you'll witness the growth of your portfolio and your knowledge base!

Other methods for developing your portfolio

In addition to the previous points, consider these methods for expanding your portfolio:

- *Competitions*: Sites and organizations such as Kaggle, DataHack, DataCamp, Data Science Dojo, the Data Science Global Impact Challenge, and DataKind host data science competitions (also called *hackathons*) where you can apply your skills to complex problems, often alongside other learners of various skill levels. These projects can add depth to your portfolio and show you can perform under pressure.

- *Volunteer work*: Nonprofits and small businesses often need data analysis but lack resources. DataKind, Data for Good, and Statistics Without Borders are just a few organizations that consistently seek data science volunteers. Volunteering your skills can result in meaningful projects for your portfolio.

- *Blogging*: Writing about your projects, explaining the methodologies used, and discussing your results can demonstrate your communication skills and ability to translate technical concepts into plain language.

Presenting your portfolio

Once you have your projects, it's crucial to present them effectively. Here's how:

- *Choose a platform*: GitHub is popular for hosting data science projects. You can include code, datasets, and documentation. Other options include Kaggle, personal websites (e.g., Canva), or blog platforms such as Medium and Towards Data Science.

- *Documentation*: Ensure each project is well documented. Include an overview of the project, the techniques used, and a discussion of the results. Clear, concise explanations are key. Documentation can be provided directly in the code using comments, or can incorporate other methods such as a README.txt file or by using **Markdown**, a markup language used for creating legible and engaging text documents. Thanks to the explosion of generative AI, there are even platforms such as Docify AI and Mintify, which autogenerate documentation from code.

- *Accessibility*: Make sure that your code is accessible, reproducible, and easy to read. Good practices include commenting your code, formatting your code, using clear variable names, organizing your code neatly, and following general coding conventions and best practices when available. Some **integrated development environments** (**IDEs**) have features that make creating accessible projects easier. For example, VS Code is an IDE that offers the Integrated Accessibility Checker and a host of extension applications that have accessibility features.

- *Visualizations*: Effective data visualizations can make your projects stand out. They can demonstrate your ability to tell data stories and present data in a meaningful, interesting, and accessible way. We will discuss visualizations and data storytelling later!

- *Create a video*: You can create a video of some of your work portfolio explaining and presenting your information. If you can create a compelling and engaging video, you can post it on YouTube and share it through your social media channels. This is another way for recruiters to see you and help you stand out. A compelling data story can encourage others to share your video, and since it is a video, it can make its way around the web, promoting you all the time, 24/7!

In conclusion, a well-crafted data science portfolio can greatly enhance your job prospects. By showcasing a range of projects demonstrating your skills, passion, and learning journey, you can make a strong impression on potential employers and set yourself apart in the competitive data science job market.

Applying for jobs

The job application process can often feel like a daunting maze, but you can navigate it effectively with the right strategies and understanding. This section outlines a range of approaches to ensure that you're not just applying for jobs but doing so strategically.

When to apply

Timeliness is a critical factor for job applications. Generally, the earlier you apply after a job is posted, the better. Employers often start reviewing applications soon after posting a job and may even begin interviewing before the application deadline. Therefore, focusing on jobs posted within the past week can increase your chances of your application being seen.

Applying in numbers

Remember, job applications are a numbers game. The more roles you apply for, the higher your chances of landing an interview. However, this doesn't mean you should apply indiscriminately. Aim for a balance between quantity and quality. Each application should be well-researched and tailored to the specific role and company.

To manage a high volume of applications, consider setting application goals per week. This approach can help you stay organized, motivated, and consistent in your job search. It's also beneficial to track your applications in a spreadsheet, noting details such as the company name, role, date of application, and any follow-up actions. This helps you stay organized and makes the process less overwhelming.

The key to a successful job search is persistence, patience, and strategy. By understanding the job market dynamics and applying these strategic application tactics, you can maximize your opportunities and land your desired data science role.

The Job Offer Funnel

# of Applications

For example, you complete 200 applications.

# of First Round Interviews

Evidence suggests that an average of approx. 8.4% progress to the 1st round of interviews.

# of Additional Interview Rounds

Typically, only a few candidates make it this far (about 2 or 3).

Job Offer

Evidence suggests an average of 36.2% will receive an offer.

Figure 2.1: The job offer funnel [4]

Writing a compelling cover letter

A cover letter allows you to elaborate on the information in your resume and show why you're a good fit for the role. Here are some tips for writing a compelling cover letter:

- *Showcase your interest*: Show that you're genuinely interested in the role and the company. Mention specific aspects of the job or company that excite you.

- *Tell your story*: Use the cover letter to tell a cohesive story about your career journey, highlighting the experiences and skills that make you a strong candidate as dictated by the job description. It's also advised to keep a record of your cover letters. Many roles with similar descriptions will result in a similar cover letter and can serve as a template for future applications.

- *Align with the company*: Show how your values, goals, or experiences align with the company's mission or culture.

- *Call to action*: End your cover letter with a call to action, expressing your interest in an interview or further discussions.

Cover letter rules

Cover letters can be a contentious topic among job seekers and recruiters. Should you always include one? The answer largely depends on the specific situation. If a job description explicitly requires or prefers a cover letter, you should certainly include one. Moreover, if you're particularly excited about a job or your resume doesn't directly align with the role, a cover letter is an excellent opportunity to express your enthusiasm and explain how your skills and experiences make you a suitable candidate.

However, crafting a compelling cover letter can take time, so it's advisable to be selective and focus on those applications where a cover letter could make a significant difference. Furthermore, you should utilize AI applications such as ChatGPT, Ramped, or CoverDoc.ai to automate as much of the writing process as possible. Websites such as Canva also provide various professional cover letter templates.

At this point, you've learned about the intricacies of the job search process. But in reality, you've only skimmed the surface – applying for data science jobs is a challenging task, especially given the various rules and best practices to consider. Luckily, like anything else, the process will become easier over time. As a result, the time it takes to apply for a job, screen job descriptions, and write introductory emails will be reduced and you will become more efficient. Before you know it, your job search will become a well-oiled machine with a growing project portfolio, a repository of cover letter templates, a refined networking schedule and strategy, and a healthy income of job alerts.

However, a job application is incomplete without a resume! In the next section, we will discuss the importance of a resume, and how to create one that will attract employers and stand out from the stack.

Constructing the Golden Resume

Your **resume** is arguably the single most important document in your job search journey. It acts as a first impression, a summary of your skills and experiences, and, ultimately, the key to unlocking the door to the interview stage. Given its importance, investing time and effort into crafting a compelling resume is critical.

In today's digital age, the initial review of your resume is often not performed by a human but rather by an algorithmic system known as an **Applicant Tracking System** (**ATS**). These systems perform the initial screening and filtering of resumes. However, while optimizing your resume for ATS, it's equally crucial to make it compelling for human readers. This is why resumes should contain a cohesive and concise structure and format.

The perfect resume myth

When creating a resume, many job seekers fall into the trap of striving for an elusive *perfect* document. They spend countless hours fine-tuning every word and agonizing over minute details. However, the truth is there is no such thing as a perfect resume. What works for one recruiter or hiring manager may not work for another, and what lands you an interview at one company may not have the same effect at another. The key to an effective resume lies not in perfection but in adaptability and relevance.

A resume is not a static document but rather a dynamic one that should be adjusted and tailored for each job application. Your goal should be to create a solid, well-structured baseline resume that effectively communicates your skills, experiences, and accomplishments. This **baseline resume** serves as a foundation that you can modify based on the specific requirements and preferences of each job you apply for.

Remember, the primary purpose of your resume is to communicate the most essential information about your qualifications for a specific role. It should provide a clear, concise, and compelling snapshot of your professional identity. It is essentially a marketing document for your professional value.

Understanding automated resume screening

ATS systems automatically scan and sort resumes, filtering out those not meeting specific criteria. Recruiters and hiring managers look for clear, concise, and well-organized resumes that effectively communicate a candidate's qualifications and potential. Therefore, your resume should strike a balance between being ATS-friendly and human-friendly.

Getting past the stack of applicant resumes and ATSs is a challenging task, but not impossible. You just have to master the guidelines and standards of resume building including formatting, terminology, and how resumes are screened. Hence, our goal is to build the best-matched resume, not the perfect resume.

Since most resumes are initially reviewed by these automated tools, not humans, understanding how an ATS works is critical to ensuring your resume passes this initial screening:

- *Keyword matching*: An ATS often screens for specific keywords related to the job description. Use websites such as `www.jobscan.co`, `resumeworded.com`, and `rampedcareers.com` to compare your resume with the job description and align your language with the terms and phrases used by the company. The job description is an excellent information source for knowing what keywords to use. Look for job-specific words repeated within the description, such as "neural networks" or "Python." You will want to ensure your resume highlights your experience using these terms.

- *Formatting*: An ATS may struggle with complex formatting. Use an ATS resume template. Keep your resume layout simple and clean, and avoid graphics, tables, columns, or unusual fonts. We will review this in the next section.

Crafting an effective resume

Rather than striving for the elusive perfection, focus on these key aspects:

- *Relevance*: Highlight the experiences, skills, and accomplishments most relevant to the job you're applying for. This goes beyond technical tools and tasks, but may also include industry terminology or areas of expertise. Use the job description as a guide to understand what the employer values most. You should address as many job description requirements and contexts as possible by incorporating them in your resume's job tasks and professional summary

- *Clarity and conciseness*: Avoid jargon and write in clear, concise language. Your goal is to make it easy for the reader to quickly grasp your qualifications. When possible, use industry-specific language to highlight your exposure and familiarity with key aspects of the job.

- *Quantifiable achievements*: Where possible, quantify your achievements. This adds credibility to your claims and makes your accomplishments more tangible. A useful framework to use while incorporating quantified achievements is the **Specific, Measurable, Achievable, Relevant, and Time-bound (SMART)** framework.

- *ATS optimization*: Include keywords and phrases from the job description to optimize your resume for the ATS. These days, we are fortunate enough to have AI tools such as Talentprise, Pyjama Jobs, and Fortay to *create* your own matching job program; these flag jobs that specifically meet your skill set based on your experience and background. Other platforms, such as Jobscan, grade your resume across many metrics such as matchability, searchability, word count, and words to avoid. Take that, ATS!

- *Proofreading*: Ensure your resume is free of spelling, grammar, and formatting errors. Mistakes can create a negative impression and suggest a lack of attention to detail.

Remember to approach your resume as a work in progress – continually seek feedback and be open to critiques. However, resist the urge to endlessly tweak your resume in search of perfection. A good resume can get you in the door, but your skills, experiences, and interview performance will ultimately land you the job. Instead of investing excessive hours perfecting your resume, spend that time improving your skills, networking, preparing for interviews, and applying for jobs. Balance is key in the job search process.

Here are some tips for crafting a resume that can impress both the ATS and human reviewers:

- *Use specific language*: Be specific in describing your skills and experiences. Instead of saying you have *experience with data analysis*, mention the specific tools, techniques, or projects you've worked on.

- *Active language*: Use action verbs to describe your responsibilities and achievements. Words such as *developed*, *analyzed*, and *implemented* can make your experiences sound more dynamic.

- *Quantifiable achievements*: Where possible, quantify your achievements – for example, *improved model accuracy by 20%* or *reduced processing time by 30*

Formatting and organization

Formatting and organizing your resume may seem straightforward; however, this section will highlight a few important reminders and some tips for data scientists. We will start by discussing some formatting reminders to give you a better chance of making it past the ATS screening process, while maintaining visual appeal for a hiring manager to review. Then, we will provide you with reminders and suggestions on organizing your resume.

First, ATSs analyze resumes for keywords and phrases that match a specific job description. However, these systems can only parse and understand your resume if correctly formatted. Here are some guidelines:

- *File type*: Save your resume as a `.docx` or `.pdf` file. These formats are the most compatible with ATSs.

- *Styles*: Refrain from adopting highly artistic or stylized resume templates, especially those that use page space inefficiently, feature too many icons or photos, or provide self-graded scales of skill aptitude. While these are aesthetically appealing, most of these features work against you. Not only do they not pass the ATS but they may take up valuable real estate on your resume. Instead, stick to formats that are well-tested and validated over time.

> **Tip**
> The Harvard Extension School publishes a *Resumes and Cover Letters* packet (available online) that provides highly effective resume formats, as well as some tips and advice to use. You can check them out here: `https://hwpi.harvard.edu/files/ocs/files/hes-resume-cover-letter-guide.pdf`.

- *Fonts*: Use standard, ATS-friendly fonts such as Arial, Helvetica, or Calibri. Avoid fancy or decorative fonts, which can confuse the ATS.

- *Font size*: Keep the font size between 10 and 12 points for easy readability.

- *Bullet points*: Use bullet points to list your skills, experiences, and accomplishments. Avoid using complicated symbols or graphics, as these can be difficult for the ATS to interpret. Although some academic positions that require resumes may be an exception, you should mainly avoid exceeding four bullet points per job (unless you have very few roles to speak of).

- *Avoid images, headers, and footers*: ATSs often struggle to read the information in images, headers, and footers, so it's best to avoid these.

- *Length*: This debate is yet to be settled, but there are tons of websites, blogs, and editorials with their own positions on resume length. In reality, it varies. If you are an entry-level employee with less than five years of experience, try keeping your resume to one page. Otherwise, the length of your resume is debatable. Remember, the goal is to write a concise, cohesive document that highlights your most applicable skills and experience. On average, a recruiter spends about 7 seconds on each resume. Thus, a resume that is too lengthy risks looking unprofessional at best, and hides the most relevant information from the recruiter at worst.

The organization of your resume should depend on your unique career history and the specific job you're applying for. However, a standard resume typically includes the following sections:

- **Contact Information**: Include your name, email address, and phone number at the top of your resume.

- **Objective** *or* **Summary**: A brief statement summarizing your career goals and qualifications. This should be tailored to each job.

- **Skills**: For a data scientist, this is an important section. This section is comprised of your hard and soft skills. Technical skills relevant to the job consist of programming languages, big data frameworks, business intelligence platforms, cloud computing platforms, IDEs, program management programs, and word processing programs, whereas soft skills typically consist of critical thinking, communication, or problem-solving skills. While you can list soft skills if space permits, these skills are often best displayed during the interview.

- **Professional History**: A reverse chronological listing of your past jobs, including your title, the company name and location, the dates of employment, and bullet points outlining your responsibilities and achievements. Refrain from listing irrelevant experiences. This may also include relevant internships and fellowships, particularly if you do not have applicable full-time experience. Try not to add more than 3-4 bullets per position.

- **Education**: A brief overview of your academic qualifications, including the degree earned, grade point average, the institution's name, and the graduation date. You can also use this section to highlight any technical certifications you have achieved or data science competitions you have competed in. If you feel your experience does not best summarize your skill set, you may also include a listing of relevant coursework here.

- **Projects**: If you are an early-career applicant or lack relevant on-the-job experience, consider including a **Projects** section that highlights some of your most relevant projects.

For recent graduates or those with less work experience, placing the **Skills** and **Education** sections near the top is advisable. However, if you have extensive work experience, prioritizing your **Professional History** section is more beneficial.

Using the correct terminology

A resume is not just a list of past jobs and education. It's a strategic document designed to market your skills and experiences to potential employers. Therefore, your terminology can significantly impact how employers perceive your qualifications and fit for the role. Additionally, using industry-specific terms, metrics, and phrases can put you ahead of the competition by flexing your familiarity with the business. This section will explore three fundamental principles of effective resume language: specificity, activeness, and quantifiability.

Specificity over generality

Specific language helps paint a vivid picture of your skills, experiences, and accomplishments, and by providing concrete examples, you will be able to demonstrate your qualifications better.

Consider these two statements:

- General: *Experienced in data analysis*
- Specific: *Leveraged Python and R to analyze a dataset of over 1 million records, identifying key trends and insights*

The first statement is too broad, but the second statement provides much more information and gives the employer a clearer understanding of your capabilities.

Active language over passive language

Active language makes your resume more dynamic and engaging. One way of gaining an active voice is by using action verbs to describe your experiences and accomplishments.

Consider these two statements:

- Passive: *A data visualization project was completed*
- Active: *Completed a data visualization project using Tableau to present complex data in an easily understandable format*

Compared to the passive statement, the active statement is more engaging and clearly communicates your role and contributions.

The Harvard Extension School, mentioned previously, also provides a helpful list of action verbs, categorized by skill areas such as leadership, communication, and technical skills – these are definitely worth checking out.

Quantifiable, fact-based language

Whenever possible, quantify your accomplishments. This adds credibility to your claims and helps employers understand the scope and impact of your work.

Consider these two statements:

- Non-quantifiable: *Improved sales by optimizing a pricing strategy*
- Quantifiable: *Improved sales by 20% in Q2 2023 by optimizing a pricing strategy, resulting in an additional revenue of $200K*

As you can see, the quantifiable statement provides a clearer picture of the impact of your work.

When working on your resume, an effective trick for structuring your accomplishments is the *Action-Problem-Result* format. This format describes an action you took to address a problem, followed by the result of your action.

Here is an example: *Implemented a new machine learning algorithm to address the issue of high churn rate, resulting in a 15% decrease in customer attrition within six months.*

The language you use in your resume can significantly impact its effectiveness. You can create a compelling document that effectively communicates your qualifications and potential by focusing on specificity, active language, and quantifiable, fact-based statements.

Industry jargon

There are some cases where using technical jargon is appropriate, and that's when you use it to show off your business knowledge. For example, becoming a data scientist in the digital marketing space means that you should have experience with optimizing industry-specific **key performance indicators (KPIs)**. These KPIs will look different for a data scientist who works in supply chain operations. Thus, including digital marketing-specific metrics such as **click-through rate (CTR)** or **return on advertising spend (ROAS)** on your resume will signal to recruiters that you have the industry-specific experience that they seek. Use these opportunities to shine!

For example, "*Designed, validated, and optimized an MMM to optimize ROAS by increasing branded search investments by 20%*" is a sentence that only digital marketers would understand, and in this case, it's a good thing.

To conclude, using the right terminology not only sells your accomplishments to hiring managers and recruiters but also signals that you have the right exposure, experience, or familiarity with the right terms of the business. This is advantageous because it shows that you already *speak the language* of the job, and hints that you will require less training. Furthermore, it is not enough to simply state tasks that you've completed on the job; you have to convey what SMART goals you've accomplished. Remember, recruiters are looking for an accomplisher, not a doer!

If you manage to say all the right things on your resume, though, you will increase your chances of getting a first-round interview.

Prepping for landing the interview

The guidelines in this section will help you increase your chances of reaching the initial screening stage and landing an interview.

Thorough interview preparation is paramount. It requires keeping pace with industry changes, researching target companies and hiring managers, and cultivating your professional brand and network.

Staying current with the fast-paced changes in the data science industry is crucial for differentiating yourself against candidates with outdated skills. Demonstrating your knowledge of the latest trends and technologies showcases your dedication and ability to master emerging challenges. Additionally, conducting extensive research on companies and hiring managers allows you to effectively align your skills and values with their needs and tailor your application and interview responses. Meanwhile, investing time in building your professional brand and network enhances your visibility and provides valuable connections and opportunities within the data science community.

By mastering these interconnected strategies, you'll position yourself for interview success and increase your chances of landing a data science job. As we've learned, landing the job goes beyond a professional resume, cover letter, and successful interview. The landscape of technology is always changing, which makes data science one of the most dynamic and exciting fields. However, it also means staying current on the latest trends and tools in the industry.

The next section will provide you with some tips on how to do just that. By the end of the section, you will be able to craft a custom skilling-up strategy to ensure your skills remain relevant and fresh.

Moore's Law

The pace of technological change can often feel akin to **Moore's Law**, the principle that the speed and capability of computers can be expected to double approximately every two years. This idea is driven by technological advancements and the exponential growth in computing power, and is a fitting metaphor for the ever-evolving tech industry. Ultimately, the challenge lies in perpetually learning, unlearning, and relearning.

As a data scientist, a significant part of your job will be to stay abreast of new developments, be it a new programming language, a revolutionary machine learning algorithm, or the latest data management system. Fortunately, there are several strategies that you can adopt to keep pace:

- *Blogs/newsletters/podcasts*: Consider subscribing to relevant data science blogs, newsletters, and podcasts. These resources can provide timely updates on the latest trends and breakthroughs. For example, *The Analytics Power Hour* is a fun and insightful podcast on the analytics profession, hosted by three analysts. *DataFramed* by DataCamp, and *Not So Standard Deviations* are also engaging and thought-provoking programs. *Medium* and *Towards Data Science* are recommended too.

- *Participate in online communities*: Participate in online communities and forums, such as GitHub, Stack Overflow, or Kaggle. These platforms offer a wealth of shared knowledge and resources and foster active discussions about the latest developments in the field. You can also find a variety of relevant social groups on LinkedIn, Discord, Slack, Meetup, and even Facebook. Plus, there are programming language-specific groups such as R-Ladies.

- *Attend conferences and workshops*: Attending conferences, webinars, and workshops can also be an effective way to learn about new tools and techniques and to network with other professionals in the field. In addition, these events often showcase the latest research and developments in the field and provide opportunities for networking with industry professionals, academics, and researchers. The **Open Data Science Conference (ODSC)**, **PyData, Data Science Summit, Rev4, Data Science Salon**, and the annual meeting for the **Institute for Operations Research and the Management Sciences (INFORMS)** are some of the most popular.

- *Online courses*: As previously mentioned, continuous learning through online courses is a great way to supplement your knowledge, especially on specialty topics. DataCamp, edX, Coursera, SoloLearn, Udacity, Udemy, Khan Academy, and CodeAcademy are examples of online course sites. Some also offer graduate degrees.

- *Review research papers*: Reading research papers and even pursuing advanced degrees can contribute significantly to keeping your skills and knowledge fresh. One of the most accessible search engines for research papers is **Google Scholar.**

Some job searches can take months to find the proper data science role; however, it is essential to do what you can to stay up to date on the state of the field. Remember, the key is a commitment to ongoing learning and curiosity about new developments. As the field continues to evolve rapidly, these strategies will help ensure that you remain at the forefront of knowledge and skill.

And as a data scientist, your learning journey never truly ends – it merely evolves.

Research, research, research

Successful interviews often hinge on preparation, which entails researching the company and the hiring manager, anticipating likely questions, and preparing for technical queries.

Researching the company

Understanding the company that you're interviewing with is crucial. This research shows respect for the company and interviewer and gives you a chance to genuinely decide whether it's the right place for you. It also makes you a prepared candidate who appears informed about the company. Here's how to approach this:

- *Company website*: The company's official website is your first and most direct source of information. Here you can understand the company's mission, products, services, goals, challenges, initiatives, organizational structure, and culture.

- *Recent news*: Look for recent news about the company to fuel your interview conversations and prove that you've done your research. This could include new product launches, partnerships, or leadership changes, as well as recent related legislation or company acquisitions.

- *LinkedIn, Glassdoor, Fishbowl, or Blind*: These platforms can provide insights into the company's culture, values, salary ranges, and employee experiences.

- *Industry trends*: Understanding the broader industry context can help you ask insightful questions and show that you're in touch with current trends.

Researching the hiring manager

Understanding the person who might hire you, the hiring manager, can be a significant advantage in your job search journey. Typically, job seekers don't get to know the hiring manager until they interview with them, however, if you are fortunate enough to find out who the hiring manager is beforehand, this opens up an opportunity to make a more interesting connection when you meet.

LinkedIn is a great resource to learn about the hiring manager's professional background. By examining their profile, you can gain insights into their career trajectory, their roles and responsibilities, and perhaps most importantly, their interests and the problems they are passionate about solving. Knowing the hiring manager's areas of interest can provide you with valuable context about what they find important in a candidate.

For example, if the hiring manager's LinkedIn profile indicates a strong interest in machine learning and AI, then during the interview, you could emphasize your skills, experiences, and projects related to these areas. This could help establish a connection with the hiring manager and demonstrate that your skills align with their interests and needs. Additionally, if you discover that you and the hiring manager have the same alma mater, what a great opportunity to connect!

Researching the hiring manager also allows you to understand the types of problems they might be hiring someone to solve. You'll be in a strong position during the interview if you can showcase how your skills and experiences make you an excellent candidate to address these problems.

However, while it's beneficial to understand the hiring manager's background, respecting their privacy is also important. Therefore, always approach this research with professionalism and respect.

If you have the opportunity to research the hiring manager, take it. It provides valuable insights that can help you tailor your interview responses and demonstrate your potential to meet their needs. It's one of the many ways to prepare for and increase your chances of landing the job.

Branding

As with any other professional field, the strength of your personal brand can be as vital as your technical skills. Your **personal brand** is the perception that others have of you based on your skills, experiences, and personal qualities.

A professional brand starts with self-awareness – you need to understand your strengths, areas of expertise, values, passions, and what differentiates you from other data scientists. Once you have a clear understanding of your unique qualities, you can start communicating this to others.

Here are a few steps to developing a professional brand:

- *Craft a consistent message*: Your resume, LinkedIn profile, and personal website (if you have one) should tell a consistent story about your skills, experiences, and career goals.

- *Showcase your work*: Whether it's a data science project that you've completed, a blog post that you've written, or a talk that you've given, make sure that your work is known by others. This helps to establish your credibility and showcase your expertise.

- *Build an online presence*: Social media platforms, especially LinkedIn, provide a great opportunity to establish a professional brand. Regularly share and engage with relevant content, showcasing your knowledge in your field

Summary

We reviewed a lot in this chapter, and you're probably overwhelmed with all of this information. But now, you should feel equipped to start your data science job search.

To begin, you were introduced to preparing and starting a job search, including how to mentally prepare for the process, and how to leverage job board sites to find leads, use them for networking, and gain insights into specific industries. Additionally, we discussed how to start to build a work portfolio to be used during your interviewing process.

Then, we looked into the other key element for your job search: the resume. Here, we discussed tips on how to craft and organize your resume to not only be noticed by someone but also to get past the applicant tracking systems that often perform the first filtering of resumes.

After that, we discussed prepping for the interview by conducting research on key companies who are hiring and staying up to date on key trends in the industry. Finally, we closed the chapter by discussing the importance of developing your personal, and professional brand, and how to do it.

By using this guide on staying current with evolving technology, being diligent about networking, and building a smart and streamlined strategy for developing resume, portfolio, and cover letter content, you can maximize your chances of landing a data science interview.

Additionally, you will remain active and adaptive to new opportunities as they arise, whether from job board sites or networking conversations. As French scientist Louis Pasteur once stated, "*Chance favors the prepared mind*." If the tools and tips from this chapter are properly leveraged, you will have the optimal opportunity to take full advantage of the chances that come your way.

In the next chapter, we will focus on helping you prepare for the technical portion of the data science interview by first looking at Python.

References

- [1] *Working Whole: How to Unite Your Career and Your Work To Live Fullfilled* by *Kourtney Whitehead* (*Simply Service, 2019*)

- [2] *The Biggest Job Search Myth, Debunked* by *Jennifer Parris*, from *Flexjobs*: `https://www.flexjobs.com/blog/post/biggest-job-search-myth-debunked/#:~:text=About%2070-80%20percent%20of%20job%20listings%20are%20never,public.%20Instead%2C%20they%E2%80%99re%20filled%20through%20word-of-mouth%2C%20or%20networking`

- [3] *Use the 70% Principle To Find Your Next Job*, by *Kelly Studer*, from *Ivy Exec*: `https://ivyexec.com/career-advice/2014/use-70-principle-find-next-job/`

- [4] *7 Benchmark Metrics to Improve Your Recruiting Funnel* by *Stephanie Sparks*, from *Jobvite*: `https://www.jobvite.com/blog/recruiting-funnel/`

Part 2: Manipulating and Managing Data

The second part of this book covers the most common coding, data wrangling, and productivity skills found in most data science jobs and interviews. From foundational to advanced concepts, this includes an introduction to essential skills in Python, data visualization, SQL, command-line scripts, and version control.

This part includes the following chapters:

- *Chapter 3, Programming with Python*
- *Chapter 4, Visualizing Data and Data Storytelling*
- *Chapter 5, Querying Databases with SQL*
- *Chapter 6, Scripting with Bash and Shell Commands in Linux*
- *Chapter 7, Using Git for Version Control*

3

Programming with Python

Starting from this chapter, we will now transition into preparing you for the technical portion of data science job interviews. For this reason, this second part of the book is best used as a study/quick reference guide as you prepare for your interviews. Therefore, feel free to skip or review chapters according to your studying needs.

In each of the following chapters, we will review key concepts and provide sample problems. Thus, it is important that you are at least familiar with introductory programming concepts, preferably with functional programming. This includes, but is not limited to, syntax, data types, variables and assignments, control flow, and packages such as pandas and numpy for data wrangling.

By the end of this chapter in particular, you will have a handle on expected Python questions within a data science interview, and know how to tackle them logically. Additionally, you will be more comfortable and confident with thinking through questions relating to control flow, variables, data types, user functions, and general data wrangling.

In this chapter, we will cover the following topics:

- Using variables, data types, and data structures
- Indexing in Python
- Using string operations
- Using Python control statements, loops, and list comprehension
- Using user-defined functions
- Handling files in Python
- Wrangling data with pandas

Using variables, data types, and data structures

In Python, **variables** are the building blocks of any code. It's simply a value of some given type assigned to an object. For example, if I set a variable called *x* equal to *10*, the variable *x* now holds that value (until it is changed). In short, variables are used to store data. Unlike some other programming languages, such as Java, the variable type does not need explicit declaration in Python. The declaration or type of a variable is determined automatically when you assign a value to it (although you can and should change data types as needed). There are several built-in data types in Python. Here are some common ones:

- **Numeric types**: There are numerous types of numeric data types, including int (integers), float (floating-point numbers), and complex (complex numbers). Numeric variables in Python are used to store numerical data:

  - **Integers** represent whole numbers without any fractional or decimal part. They can be positive or negative. In Python, integers are represented by the int type. Take the following example:

    ```
    x = 5
    print(type(x))  # <class 'int'>
    ```

  - **Floats (floating-point numbers)** represent numbers with fractional or decimal parts. They can be positive or negative. In Python, floats are represented by the float type. Take the following example:

    ```
    y = 5.5
    print(type(y))  # <class 'float'>
    ```

  - **Complex numbers** represent numbers with both real and imaginary parts. They are written in the form $a + bj$, where a represents the real part and b represents the imaginary part. In Python, complex numbers are represented by the complex type. The imaginary part is denoted using the imaginary unit j or J. Take the following example:

    ```
    z = 1+2j
    print(type(z))  # <class 'complex'>
    ```

- **Sequence types** are data types that represent an ordered collection of elements, which can be from various data types. Thus, they allow you to store multiple items in a single object and access them by their position or index within the sequence. For example, these may include str (strings), list (lists), and tuple (tuples):

  ```
  # strings
  s = 'Hello, World!'
  print(type(s))  # <class 'str'>

  # lists
  l = [1, 2, 3, 4, 5]
  print(type(l))  # <class 'list'>
  ```

```
# tuples
t = (1, 2, 3, 4, 5)
print(type(t))  # <class 'tuple'>
```

Tuples may seem similar to lists, and indeed they are. However, there are some key differences. Perhaps one of the most important differences is immutability – tuples are immutable, whereas items in lists can be changed after the list is created. Additionally, you may note that tuples utilize parentheses as opposed to brackets.

> **Note**
>
> Lists are generally used when the order and the ability to modify the elements are important. They are commonly used for dynamic data where the size or contents may change over time. Tuples, on the other hand, being immutable, are often used when you want to ensure that the collection of elements remains unchanged. Tuples are also used for situations where you want to enforce that the elements are not modified.

- The **Boolean type** in Python represents True or False values, which may also be represented by the integers 1 and 0, respectively. These values are used to perform logical operations and control the flow of programs based on conditions:

```
# boolean
b = True
a = False
print(type(b))  # <class 'bool'>
print(type(a))  # <class 'bool'>
```

- **Dictionaries** are mutable mapping types that store data in key-value pairs. Each key in a dictionary must be unique, and it is used to access its corresponding value. Dictionaries are defined using curly braces ({ }) or the dict() constructor, with key-value pairs separated by colons (:). Take the following example:

```
# dictionary
d = {'name': 'John', 'age': 30}
print(type(d))  # <class 'dict'>
```

- **None type**: This data type has a single value, None:

```
# None
n = None
print(type(n))  # <class 'NoneType'>
```

- **DataFrame**: This is a two-dimensional, tabular data structure commonly used in structured databases and data analysis. Because of the ease that DataFrames provide for data manipulation, it has become a standard data structure in analytics, or any role that requires significant data wrangling and preparation.

  The functionality perks include simple indexing, filtering, sorting, aggregating, and calculations. Dataframes also offer convenient methods for importing and exporting data from various file formats, such as CSV, Excel, or SQL databases.

  A DataFrame consists of two dimensions: columns and rows. Each column represents a variable or feature that describes an attribute or characteristic of the row; the row represents an observation or record:

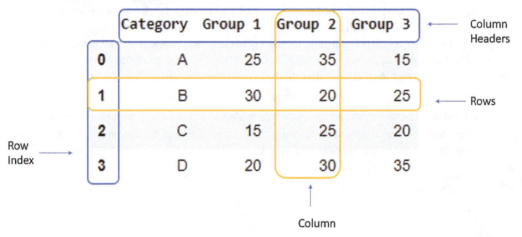

Figure 3.1: Dataframe example

> **Note**
>
> You'll see many of these terms used interchangeably throughout the book. Keep in mind that *Row = Record = Observation* and *Column = Field = Feature = Attribute*.

After that discussion of types, it is also important to note that Python is a dynamically typed language, which means that the variable type can change during the execution of a program. See this example:

```
var = 10
print(type(var))    # <class 'int'>

var = 'Hello'
print(type(var))    # <class 'str'>
```

In this example, `var` is first an integer, then it becomes a string. In other words, Python allows the re-declaration of a mutable variable.

Assessment

Consider the following Python code:

```python
x = 100

def my_func():
    x = [10, 20, 30]
    print('x inside function:', x)

my_func()
print('x outside function:', x)
```

Now, answer the following questions:

1. What is the data type of x inside the function, and what is its scope?

2. What is the data type of x outside the function, and what is its scope?

3. What will the output of the code be?

> **Note**
>
> This chapter will test you on the concepts that you learn. A great way to tackle interview questions is by using the *G.U.E.S.S method*. And no, this doesn't mean just guessing! **G.U.E.S.S** is an acronym for **Given, Unknown, Equation, Solve, Solution**. This method is typically taught with math (as you can guess by the term *equation*), but it's great for coding as well, particularly when working on multi-step and/or complex problems. The method implores problem-solvers to start out with given information or data, identify the unknown or problem, identify the equation (or formula or code) to tackle the problem, solve the problem, and provide the solution. Feel free to give it a try!

Answers

1. Inside the `my_func` function, x is a list. This x is local to `my_func` (it has local scope).

2. Outside the function, x is an integer. This x is in the global scope of the script.

3. The output of the code will be the following:

    ```
    x inside function: [10, 20, 30]
    x outside function: 100
    ```

In this example, the `my_func` function creates a new local variable, x, which doesn't affect the global x. Hence, the global x still has its original value after the function call.

Indexing in Python

To access values within a data object, we use indexing. **Indexing** is the process of accessing individual elements within a data structure. In this case, the data structure is a list, but as you will soon learn, indexing is applicable to many data structures.

> **Note**
> Each element or item within a data structure is assigned a unique index or position, starting from a specific value. In Python, this value is 0. This means that the first position in any data structure in Python is located at index 0, followed by the second position, which is located at index 1, and so on.

Indexing allows you to retrieve or manipulate specific elements within the data structure by specifying their index. It provides a way to refer to elements individually rather than accessing the entire data structure as a whole.

The basic syntax for indexing a list or tuple in Python is as follows:

```
list_or_tuple_name[index_position]
```

The `list_or_tuple_name` object is the name of the list and `index_position` is the position of the element you want to access. Here's an example:

```
languages = ['python', 'r', 'java', 'c', 'go']
print(languages[0]  #Output: 'python'
```

In this example, `languages[0]` retrieves the element at index 0, which is the first element, `'python'`. Similarly, `languages[2]` retrieves the element at index 2, which is `'java'`.

When it comes to indexing dictionaries, instead of indexing with integer positions, dictionaries use keys to access their corresponding values. You can use square brackets, `[]`, with the key inside to retrieve the value. Here's an example:

```
my_dict = {'name': 'John', 'age': 30, 'city': 'New York'}
print(my_dict['name'])   # 'Output: John'
print(my_dict['age'])    # Output: 30
```

Later, we will dive into indexing DataFrames when we discuss selecting data in pandas, and string indexing when we discuss string operations.

Using string operations

String operations are very common when working with Python and text data. Therefore, this section will review how to initialize a string, string indexing/slicing, and some common string methods.

> **Note**
>
> We will not review string regular expressions, as this is a large topic with significant depth. Check out *Mastering Python Regular Expressions* by Victor Romero and Felix L. Luis for more instructions on this topic.

Initializing a string

Python allows for string initialization (creation) in several ways. Two ways include single quotes (' ') and double quotes (" "):

```python
# Single quotes
s = 'Hello, World!'
print(s)  # prints: Hello, World!

# Double quotes
s = "Hello, World!"
print(s)  # prints: Hello, World!
```

Single and double quotes are basically interchangeable. The only difference comes into play when you have a quote mark (single or double) inside a string. For example, one common scenario is when you want to include quotes within a string. To achieve this, you can use one type of quote mark to define the string and the other type of quote mark within the string. Here's an example:

```python
quote = "She said 'I want ice cream!' "
```

In this example, the string is defined using double quotes, and the single quotes within the string are included as part of the string itself. You could have also done the inverse like so:

```python
quote = 'She said "I want ice cream!" '
```

For code legibility, it is recommended to be consistent, regardless of the method you use.

String indexing

In Python, strings are sequences of characters, and each character has a position or index associated with it. String indexing allows you to access individual characters in a string, while string slicing allows you to access a substring from a string.

Strings are also zero-indexed in Python. That is, the index of the first character is 0, the index of the second character is 1, and so on. Python also supports negative indexing, where the index of the last character is -1, the index of the second to last character is -2, and so on. Just take note, spaces count as a character, and negative indexing actually begins at 1.

For example, consider the following string of text assigned to the variable s. We can access each character in the string by using string indexes:

```
s = "Hello, World!"

# indexing
print(s[0])   # prints: H
print(s[7])   # prints: W
print(s[-1])  # prints: !
```

Slicing is another method of accessing string characters, and is most often used to extract a window or substring from a string. The syntax for slicing is string_variable[start:stop:step] – start is the index where the slice starts (inclusive), stop is the index where the slice ends (exclusive), and step is an optional parameter used to specify the step value (also known as the number of characters to skip). If step is negative, the slicing will begin from right to left instead of the default evaluation method (left to right).

Consider the same string object, s, as before. Suppose we want to *slice* the strings to access a window of the string as opposed to just one position within the string:

```
s = "Hello, World!"

# slicing
print(s[0:5])   # prints: Hello
print(s[7:12])  # prints: World
print(s[::2])   # prints: Hlo ol!
print(s[::-1])  # prints: !dlroW ,olleH (reverses the string)
```

Let's look at each of the slices:

- In the first slice, s[0:5], the slicing begins at index 0 and stops at index 5, so it extracts the first five characters

- In the second slice, s[7:12], it starts at index 7 and stops at index 12, so it extracts the word World

- In the third slice, s[::2], no start or stop is specified, so it goes through the entire string with a step of 2, extracting every other character

- In the last slice, s[::-1], a negative step is used to reverse the string

Python provides a variety of built-in methods for string manipulation. Here are explanations and examples of `strip()`, `split()`, `join()`, `replace()`, and `find()`:

- `strip()`: This method removes leading and trailing whitespace from a string. It's often used in data cleaning when we want to remove unwanted spaces:

```
s = "  Hello, World!  "
print(s.strip())  # prints: "Hello, World!"
```

- `split()`: This method splits a string into a list where each word is a separate element. This is extremely useful in **natural language processing** (**NLP**) tasks for tokenization and other data transformation tasks:

```
s = "Hello, World!"
print(s.split())  # prints: ['Hello,', 'World!']
print(s.split(','))  # prints: ['Hello', ' World!']
```

You can also specify a separator to split on (for example, to split a string into sentences, you might split on the period (.) character).

- `join()`: This method combines a list of strings into one string. You call this method on the string you want to use as the separator:

```
words = ['Hello', 'World!']
print(' '.join(words))  # prints: "Hello World!"
```

- `replace()`: This method replaces occurrences of a substring within a string with another substring. It's often used in data cleaning and preprocessing:

```
s = "Hello, World!"
print(s.replace('World', 'Python'))  # prints: "Hello, Python!"
```

- `find()`: This method returns the index of the first occurrence of a substring in a string. If the substring is not found, it returns -1:

```
s = "Hello, World!"
print(s.find('World'))  # prints: 7
print(s.find('Python'))  # prints: -1
```

Text mining and NLP tasks are generally beyond the scope of this book, but we recommend that you read up on it if you're specifically interested in that area of data science!

Assessment

Consider the following Python string:

```
s = "Data Science with Python"
```

Now complete the following tasks:

1. What does s[5:11] return?
2. What does s[::-1] return?
3. Use a string method to split s into individual words and store the result in a list.
4. Use a string method to convert s to lowercase.

Answers

1. s[5:11] returns the string "Scienc". It starts at index 5 (inclusive) and ends at index 11 (exclusive).
2. s[::-1] returns the reverse of the string s, that is, "nohtyP htiw ecneicS ataD".
3. The split() method can be used to split s into individual words: words = s.split(). This will give words as ['Data', 'Science', 'with', 'Python'].
4. The lower() method can be used to convert s to lowercase: lowercase_s = s.lower(). This will give lowercase_s as "data science with python".

Assessment

Consider the following Python string:

```
s = "   Hello,    World!    "
```

Now complete the following tasks:

1. Use a string method to remove the leading and trailing whitespaces.
2. Use a string method to replace "World" with "Python".
3. Use a string method to find the index of the first occurrence of "World".

Answers

1. The strip() method can be used to remove the leading and trailing whitespaces: s_stripped = s.strip(). This will give s_stripped as "Hello, World!".
2. The replace() method can be used to replace "World" with "Python": s_replaced = s.replace("World", "Python"). This will give s_replaced as " Hello, Python! ".
3. The find() method can be used to find the index of the first occurrence of "World": index = s.find("World"). This will give index as 11.

Using Python control statements, loops, and list comprehensions

Control statements are used for various tasks. For example, they're used to filter data based on certain conditions, perform a calculation on each item in a list, iterate through rows in a dataframe, and more. Additionally, list comprehensions are widely used in data science as they provide efficiency and legibility. It's often used in data cleaning and preprocessing tasks, feature engineering, and more.

Control statements in Python allow you to control the flow of your program's execution based on certain conditions or loops. The main types of control statements are conditional statements (such as `if`, `elif`, and `else`) and loop statements (such as `for` and `while`).

Meanwhile, list comprehensions are a sort of short-hand approach to writing loop statements. More specifically, they are a shorter, more concise syntax for creating a list based on the values of an existing list.

Conditional statements such as if, elif, and else

Conditional statements are probably one of the easiest control statements to understand because they operate (and are written) in a way that reflects how humans mentally evaluate *if-else* scenarios. Let us consider the `if`, `elif`, and `else` conditional statements:

- `if` is used to test a specific condition. If the condition is true, the code block within the `if` statement will be executed:

```
x = 10
if x > 0:
    print("x is positive") #Output: "x is positive"
```

- `elif`, which stands for *else if*, is used to chain multiple conditions. It's particularly handy when used after an `if` or another `elif` statement. This is because if the result of an `if` code block is false, the next condition (`elif`) will be evaluated. If the `elif` condition is evaluated as true, it will be executed. In the following example, the `if` statement is evaluated first. In this particular case, x is greater than 0; thus, the initial `if` statement is false. This prompts the program to evaluate the following `elif` statement, which is true:

```
x = 10
if x < 0:
    print("x is negative")
elif x > 0:
    print("x is positive") # Output: "x is positive"
```

- `else` is the last statement evaluated after the `if` and `elif` code blocks have been evaluated. `else` is almost identical to `elif` in functionality, but the major difference between the two is that you use `else` for the last logic statement check. `elif` is used to pass the logical check to another logic assessment. `else` is the very last logical statement to be evaluated – hence, the criteria in `else`:

```python
x = -10
if x > 0:
    print("x is positive")
elif x == 0:
    print("x is zero")
else:
    print("x is negative") # Output: "x is negative"
```

Loop statements such as for and while

Loops are another category of control statements used to evaluate a block of code iteratively.

To begin, let us consider a `for` loop example for inspiration. **for loops** are a control flow mechanism used to evaluate items in an iterable data structure. This is most useful when you want to perform an operation for multiple items in an object such as a list or string.

Imagine you have a bag of M&Ms. You are tasked with drawing one M&M at a time and evaluating whether it is an orange M&M or not. If we write this process in *pseudo-code*, it might look something like this:

```python
for M&M in bag:
    if M&M == "orange":
      print("This is orange!")
    else:
      print("Not orange")
```

The block of code within the `for` loop is executed once for each item in the object:

```python
for i in range(3):
    print(i)
# prints:
# 0
# 1
# 2
```

> **Note**
>
> `for` loops become even more powerful when combined with other control flow operations, such as `if` statements, and other useful mechanisms, such as functions. When combined, these tools allow you to perform operations, calculations, evaluations, and revisions on multiple items in an iterable object. Notice, we already snuck in an example of using `for` loops and `if` statements together in the M&M example. Did you catch it?

We also have **while loops**, which are used when you want to repeat a block of code as long as a certain condition is true. The condition is a Boolean expression that determines whether the loop should continue executing. As long as the condition evaluates to true, the code block inside the loop will execute. Once the condition becomes false, the loop will terminate. Here is an example:

```
i = 0
while i < 3:
    print(i)
    i += 1
# prints:
# 0
# 1
# 2
```

Unlike `for` loops, `while` loops iteratively evaluate a statement until it is no longer true, or if a break is inserted. In the previous example, the interpreter will loop over the statement until the object `i` is no longer less than 3.

You may be wondering: what happens if `i` is indefinitely less than 3? The answer is the program will (try to) run indefinitely. In the previous example, `i += 1` specifies that the variable will increment by a value of 1 every iteration. Without this stipulation, the code would run forever. This is where the `break` operator comes into play.

The following example demonstrates how to use breaks. In this example, we use the `break` statement to exit a `while` loop. This is a typical use case of a `break` statement, when you want to terminate the loop based on a specific condition:

```
count = 1

while True:
    print(count)
    count += 1

    if count > 5:
        break
```

The output of the code is as follows:

```
1
2
3
4
5
```

In this example, the while loop condition was set to True to create an infinite loop. However, the loop is terminated using the break statement when the count exceeds 5. This allows us to print numbers from 1 to 5 and then exit the loop.

List comprehension

As previously mentioned, a list comprehension can be thought of as a more compact and concise method of writing for loops. Here is the basic syntax of list comprehension:

```
[expression for item in iterable]
```

The expression is applied to each item in the iterable, and the results are collected into a new list.

Let's take an example of creating a list of squares for the numbers 0 to 9:

```
squares = [x**2 for x in range(10)]
print(squares)  # prints: [0, 1, 4, 9, 16, 25, 36, 49, 64, 81]
```

You can also include an if condition in a list comprehension to filter the items:

```
even_squares = [x**2 for x in range(10) if x % 2 == 0]
print(even_squares)  # prints: [0, 4, 16, 36, 64]
```

In this example, only the squares of even numbers are included in the new list.

Assessment

Consider the following Python code:

```
numbers = [5, 2, -3, 7, -1, 4]
total = 0
for number in numbers:
    if number > 0:
        total += number
print(total)
```

What value will be printed, and why?

Answer

The value printed will be 18. The `for` loop iterates over each number in the `numbers` list. If the number is positive (greater than 0), it is added to the total. Thus, the total will be the sum of all positive numbers in the `numbers` list, which is $5 + 2 + 7 + 4 = 18$.

Assessment

Write a list comprehension that will create a new list of squares that contains the squares of all numbers from 1 to 10.

Answer

The following list comprehension will create the required list:

```
squares = [x**2 for x in range(1, 11)]
```

This will produce the list `[1, 4, 9, 16, 25, 36, 49, 64, 81, 100]`, which are the squares of the numbers from 1 to 10.

Note that `range(1, 11)` is used instead of `range(1, 10)` because the stop value given to the `range` function is exclusive. Thus, to include 10 in the range, we need to specify the stop value as 11.

Using user-defined functions

Sometimes, you may need to create your own function to perform very specific operations. This is common in the data science world, especially as it relates to data cleaning, preprocessing, and modeling activities.

In this section, we will discuss **user-defined functions**, which are functions created by the programmer to perform specific tasks. They are not unlike mathematical functions, which (usually) take some inputs and (often) produce some outputs. User-defined functions are designed to take 0 or more inputs, do some specific computation(s) (we'll just call it *stuff*), and produce an output.

This process is especially helpful when performing repeated tasks. In fact, the rule of thumb is to use it if you have to do a task more than once. In more advanced cases, user functions are also helpful for code reusability, organization, readability, and maintainability.

Breaking down the user-defined function syntax

When used effectively, user-defined functions are your best friend. Like everything else in programming, functions can get pretty complex, but the fundamentals are fairly simple.

Let's take a look at the syntax:

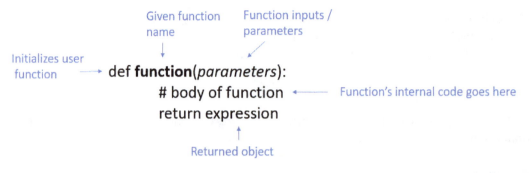

Figure 3.2: User-defined function syntax

In summary, if we were running a kitchen, the function name is the name of the receipt, the parameters are the ingredients, the statement(s) are the cooking instructions, and the return expression is the delivery method (… takeout, anyone?).

Doing "stuff" with user-defined functions

There are different types of user-defined functions. Function types in this book are based on the number of inputs:

- **No parameters**: User-defined functions with no parameters might seem peculiar at first, but sometimes, you need to do some *stuff* without additional information beyond what you describe in the body of the function. For example, consider the following function:

```
# Define a function that gives us some Vulcan wisdom
def vulcanGreeting():
    print("Live long and prosper")
#Call the function
vulcanGreeting() #Output: Live long and prosper
```

This code creates a function called `vulcanGreeting()` that prints the text *Live long and prosper* (a *Star Trek* reference).

- **One or multiple parameters**: Some functions will have at least one input parameter. This is especially true in data science, where functions are used to manipulate data. To manipulate existing data objects, an input is required. Let's look at an example:

```
# Calculate a column's average and return the value
def calculate_average(column):
    average = column.mean()
    return average
```

This code creates a function called `calculate_average`, which calculates the average (mean) of an input DataFrame column and returns the value. This function can now be applied to a DataFrame column to return its average.

But suppose you wanted to append the result to a DataFrame. This is a common need so that the results can be further explored. The following code demonstrates how to achieve this using three inputs instead of one:

```
# Calculate and append a new column "Sales" to a DataFrame that
multiplies the units and price columns
  def calculate_sales(df, units_col, price_col):
      df['Sales'] = df[units_col] * df[price_col]
      return df
```

Let's break this code down:

- **Inputs**: This function takes three parameters: `df`, `units_col`, and `price_col`. The first parameter is the DataFrame object, which contains the columns that represent the `units` and `price` columns (the other two parameters), respectively.

- **Body**: The `body` function creates a new column called `Sales`, which is calculated by multiplying the `units_col` and `price_col` column values (note: this happens for each row of the dataset).

- **Return**: The return statement returns the DataFrame, which now has the `Sales` column intact.

Note that the functionality of our function is identical to this algebraic expression: *Sales = Units x Price*. When the function is applied to the inputs, it is evaluated for each row of the dataset. Hence, every row is assigned a *sales* value in the `Sales` column.

- **Default parameters**: There are also functions that take default parameters. These are most useful in situations where you want to designate a default, static value. There are a number of scenarios where it might be advantageous to set a default setting (for example, when you want to provide a default functionality when a parameter isn't provided). Consider the following example:

```
# Write a function to greet someone by name
def greet(name="Guest"):
    greeting = "Hello, " + name + "!"
    return greeting


# Calling the function without providing an argument
default_greeting = greet()
print(default_greeting) #Output: Hello, Guest!

# Calling the function with an argument
custom_greeting = greet("Alice")
print(custom_greeting) #Output: Hello, Alice!
```

Let's dissect what the function is doing. The greet function takes an argument, but notice it's already assigned a value (in this case, "Guest"). The assigned value is the object's default value. This means that the function will always assume the default value, *unless* otherwise overwritten. Regard how the output changes when the function is called without a parameter, versus when it's called with one.

Getting familiar with lambda functions

As discussed, functions can get pretty complex, but the best functions are simple. Simple functions are further simplified by providing a simpler syntax option. Enter lambda functions!

Remember list comprehensions? They're the expedited, streamlined version of for loops. Functions have something similar, and they're called lambda functions! **Lambda functions** are used to create single-line functions in Python. Instead of using the def approach, lambda functions are defined using the lambda keyword, followed by a list of input arguments, a colon (:), and the expression or code block to be executed. Their syntax is as follows:

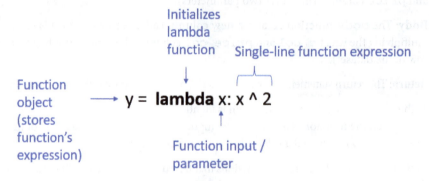

Figure 3.3: Lambda function syntax

The following code shows two different methods of accomplishing the same thing. The first one uses the user-defined function approach:

```
# Create a user-defined function that returns the sum of 2 variables
def add_numbers(a, b):
    return a + b
result = add_numbers(3, 4)
print(result)  # Output: 7
```

And the second one leverages a lambda function:

```
# Create a lambda function that returns the sum of 2 variables
add_numbers = lambda a, b: a + b

result = add_numbers(3, 4)
print(result)  # Output: 7
```

> **Note**
>
> If your lambda function takes more than one line of code, it's better to use a regular user-defined function. Furthermore, a single-line comment should suffice for documentation purposes.

Creating good functions

Here are some best practice guidelines to save some time and headaches while creating functions:

- Remember, your function name should be descriptive, but simple.
- Your function should serve a single purpose. Avoid duplication. No matter the purpose of your function or the *stuff* you have your function doing, it should only do that *stuff* once.
- Use docstrings. There are multiple docstring conventions that you can read about, such as Google's format, `reStructuredText` (`reST`), or `Numpydoc`. But as long as your docstrings adequately describe the function's functionality, parameters, and output(s), you're good.

Assessment

Let us now review some hypothetical interview questions (note, you may choose the name of the function!):

1. Write a function that calculates the area of a rectangle given its length and width. Hint: Area = length x width.
2. Write a function that returns `"Even"` if a given number is even, and returns `"Odd"` if a given number is odd.
3. Write a function that counts the number of vowels in a given string.
4. Write a function that takes a DataFrame object as input and returns the count of missing values (NaN) in each column.

Answers

1. Given the provided algebra formula, we know how to theoretically calculate the area of a rectangle. All we need is the width and length – these are our inputs!

    ```python
    def calculate_rectangle_area(length, width):
        area = length * width
        return area
    ```

    Bonus points if you wrote the answer as a lambda expression, as seen in the following code:

    ```python
    calculate_rectangle_area = lambda length, width: length * width
    ```

2. *If* is an important word in this question. It hints that we may need to use an if/else control statement. Based on the question, what conditions might we want to check? Well, we want to check whether a number is odd (condition 1) or even (condition 2). We also know that these are mutually exclusive. If a number is odd, it can't be even (and vice versa). Furthermore, *a given number* sounds a lot like an input! Thus, we have so far devised that we are writing a function with a single input (a number) and we want to check whether (using an if statement) that input is odd (condition 1) or even (condition 2).

    From our previous experience with if statements, we know that conditions are designated using if and else (or elif if there are more than two conditions). Additionally, we know what each condition must return if true (in this case, "Odd" or "Even"). All that's left is to determine the simplest way to check whether a number is odd or even:

    ```python
    def check_even_odd(number):
        if number % 2 == 0:
            return "Even"
        else:
            return "Odd"
    ```

3. In this example, we are given a string, and we want to count how many vowels it contains. How might we approach this problem? Well, the string is one input. It then sounds like we will be evaluating each character in the string. What Python control flow syntax helps us assess each index in an object? You guessed it – for loops! What other information do we need? We should probably specify what counts as a vowel (hint: the value sought out in the for loop is case sensitive!). You can see this here:

    ```python
    def count_vowels(string):
        vowels = "aeiouAEIOU"
        count = 0
        for char in string:
            if char in vowels:
                count += 1
        return count
    ```

4. This problem tells us upfront that it will take a single DataFrame object as an input. We also know that the function should return the count of missing values, even if we don't know how to derive it yet. Sometimes, it's helpful to assign a placeholder variable for the final output, even if we don't know how to calculate it just yet.

    This leaves us with a known input and output, but we still need to figure out what *stuff* needs to happen in the body of the function. For starters, we should probably assign the output value to some expression. After all, the placeholder variable will not return anything unless we assign it a value. Now, here's the tricky part – how do we count missing values? In Python, there are two useful methods: isnull() and sum(). Here is how you can string these operations together:

    ```python
    def count_missing_values(df):
        missing_counts = df.isnull().sum()
        return missing_counts
    ```

Handling files in Python

In Python, the built-in open function is used to open a file, and it returns a file object. Once a file is opened, you can read its contents using the read method. However, an important aspect to consider while managing files is ensuring they are closed after use, allowing for the setup and teardown of computational resources. One way to accomplish this is by using context managers.

Context managers are an object that manages the context of a block of code, typically with a with statement. It's particularly useful for setting up and tearing down computational resources, such as efficiently opening and closing files. In short, the with keyword, which automatically closes the file once the nested block of code is executed, is more efficient and reduces the risk of a file not being properly closed.

The syntax to open files using context managers is as follows:

```python
with open(<file_name.csv>) as file_object:
    # Code block
```

Here's a concrete example of how to open and read a file:

```python
with open('file.txt', 'r') as file:
    content = file.read()
print(content)
```

In this example, file.txt is the name of the file to open, and r is the mode in which the file should be opened. The r stands for read mode, which allows the contents of the file to be read but not modified.

The with open(...) as file: line opens the file and assigns the resulting file object to the file variable. Then, file.read() reads the contents of the file and returns it as a string, which is assigned to the content variable. After the with block is executed (even if an error occurs within the block), the file is automatically closed.

Opening files with pandas

The pandas library in Python provides high-performance, easy-to-use data manipulation and analysis tools, and is frequently used in data science roles.

One of the most commonly used pandas functions for reading in data is `read_csv()`. Here's an example of how you might use it:

```
import pandas as pd

df = pd.read_csv('file.csv')

print(df.head())  # print the first 5 rows of the data
```

In this example, the `read_csv` function reads the CSV file named `file.csv`. The resulting object is assigned to the `df` variable. The `head()` function is then used to print the first five rows of the dataframe. If you want to print the entire dataframe, you could simply write `print(df)`.

As previously mentioned, pandas also offers a function to convert a file to a DataFrame. Simply use `pd.DataFrame()` as seen in the following code:

```
# Create a DataFrame from the
df = pd.DataFrame(df)
# Print the DataFrame
print(df.head) #Outputs the first 5 rows of the DataFrame
```

Assessment

Consider the following Python code snippet:

```
with open('data.txt', 'r') as file:
    content = file.read()
print(content)
```

Now, answer the following questions:

1. What does this code do?
2. What is the significance of `r` in the open function?
3. What is the role of `with` in opening the file?

Answers

1. This code opens a file named `data.txt` in read mode (`r`), reads its entire contents into the `content` string variable, and then prints the contents. After the `with` block is executed, the `'data.txt'` file is automatically closed.

2. The `r` in the `open` function stands for `"read"`, which means the file is opened in `"read"` mode. In this mode, you can read from the file, but you can't write to or modify the file.

3. If `'data.txt'` does not exist or can't be found in the directory from which the Python script is run, Python will raise a **FileNotFoundError** message.

4. `with` is used when working with unmanaged resources (such as file streams). It's a neat bit of syntax that ensures the `File` object, `file`, is properly closed after usage. It sets up a context where the file is open, and at the end of this context, it automatically closes the file, even if exceptions were raised within the context. This makes it the best practice for resource management in Python.

Wrangling data with pandas

Data wrangling is one of the most important topics in data science interviews. For starters, data is often not presented in an analysis-ready format, which makes it necessary for data modeling preprocessing and addressing data quality concerns. Thus, data scientists can spend upward of 80% of their time cleaning and wrangling data [1].

Furthermore, data wrangling skills demonstrate your comfort and fluency with computer programming. Having the ability to use functions, loops, indexing, aggregation, filtering, and forming calculations will serve you well in your data science journey, enabling you to complete work quickly and efficiently. It is also fundamental for **extract, transform, load** (ETL) activities, querying data, data modeling, descriptive statistics, reporting, and a host of other data tasks.

In this section, we will review a couple of common data wrangling challenges, including handling missing data, filtering data, merging, and aggregating data.

Handling missing data

Sometimes, data is incomplete. Missing data is most often indicated by completely blank values, NaN values, or null values. There's a number of reasons this can happen, ranging from erroneously collected or deleted data to data that was never provided. In fact, there are even categories of missing data, which can inform how missing values are to be treated. The following are some categories of missing data:

- **Missing completely at random (MCAR):** This is data that is missing in a randomly distributed fashion across the entire variable (e.g., column, field, attribute, and feature), regardless of other variables. In other words, the data is missing completely at random, and its missingness is not correlated with other field values:

  - Example: If you have an **electronic health record** (EHR) dataset, and patient social security numbers are missing throughout that field, regardless of patient location, ethnicity, and BMI, this is MCAR.

- The simplest approach is to remove the missing data points (rows). This ensures that any analysis is not biased by the missing values. This is achievable by using pandas' `dropna()` function.

- **Missing at random (MAR)**: This is missing data that is systematic but can be explained by other observed variables in the dataset:

  - Example: For another EHR dataset, the `"Smoking Status"` field is missing for some patients, but the missingness can be explained by another observed variable, such as `"Age"`. Younger patients are less likely to have their smoking status recorded.

  - You can use methods such as mean imputation, median imputation, or predictive imputation (e.g., regression imputation) to fill in the missing values. This is achievable by using pandas' `fillna()` function. The choice of imputation method is up to the analyst, but there are some rules of thumb. Using the *mean* of the field to impute missing values is a suitable method for data with a symmetric (e.g., normal) distribution. Using the *median* of the field is more suitable for data with a skewed distribution. Advanced methods such as *regression* might be useful when there's a significant correlation between the variable with missing values and other observed variables.

- **Missing not at random (MNAR)**: This is when the missingness is related to unobserved factors or missing data itself:

  - Example: For the same EHR dataset, the `"Mental Health Diagnosis"` field is missing for some patients, but the missingness is related to the severity of their mental health condition. Patients with more severe conditions are less likely to have their diagnosis recorded.

  - MNAR is the most complicated case to remedy because the missingness is not easily explained by observed variables. Thus, it is important to carefully analyze the reasons for missingness and consider more sophisticated techniques such as multiple imputation or maximum likelihood estimation.

Let's take a closer look at an example, using the following dataframe:

```python
import pandas as pd
import numpy as np

df = pd.DataFrame({
    'A': [1, 2, np.nan],
    'B': [5, np.nan, np.nan],
    'C': [1, 2, 3]
})
```

Here are some ways you can handle missing values:

- **Drop missing values**: The `dropna()` function removes missing values. By default, it removes any row with at least one missing value:

  ```
  print(df.dropna())
  ```

- **Fill in missing values**: The `fillna()` function fills in (also called imputes) missing values with a value of your choice. Here we replace the missing data with the `'FILL VALUE'` string:

  ```
  print(df.fillna(value='FILL VALUE'))
  ```

  Then here is an example imputing missing data with the mean:

  ```
  print(df['A'].fillna(value=df['A'].mean()))
  ```

  In this case we take column `'A'` and fill in the missing values with the mean of column `'A'`.

You may also use regression to impute the data, but this is a little more involved, and we haven't covered regression yet. We will discuss regression in a later session.

Selecting data

Selecting data is a very common operation when you're working with data. With pandas, you can select data in a dataframe or Series in several different ways.

Suppose you have the following dataframe:

```
import pandas as pd

df = pd.DataFrame({
    'A': [1, 2, 3, 4],
    'B': [5, 6, 7, 8],
    'C': ['a', 'b', 'c', 'd']
})
```

Here's how you can select specific parts of the data:

- **Selecting columns**: You can select a single column using `df['ColumnName']` and multiple columns using `df[['Column1', 'Column2']]`:

  ```
  # select column 'A'
  print(df['A'])

  # select column 'A' and 'B'
  print(df[['A', 'B']])
  ```

- **Selecting rows**: You can use slicing to select rows just like you would with a list:

```
# select the first 2 rows
print(df[0:2])
```

- **Selecting by condition**: This is where pandas really shines. You can quickly filter rows based on the values in one or more columns:

```
# select rows where 'A' is greater than 2
print(df[df['A'] > 2])

# select rows where 'A' is greater than 2 and 'B' is less than 8
print(df[(df['A'] > 2) & (df['B'] < 8)])
```

- **Filter with the query method**: This allows you to filter using a string expression:

```
# select rows where 'A' is greater than 'B'
print(df.query('A > B'))
```

- **Selecting with loc() and iloc()**: There is also another school of thought for data selection, provided by the pandas package. The loc() and iloc() indexing methods are specific to pandas dataframes. They are designed to provide a convenient way to select and access specific rows and columns of a dataframe based on their labels or integer positions, respectively. Here are some notable differences between the two:

  - loc(): This method allows data selection based on column labels and/or row indices to identify and retrieve data.

  - iloc(): This method allows selection based on integer positions of rows and columns to locate and retrieve data. Note, it uses exclusive slicing, meaning that the stop index is not included in the selection. It also supports position-based slicing and indexing.

Both loc() and iloc() follow a similar syntax. You can see the loc() syntax here:

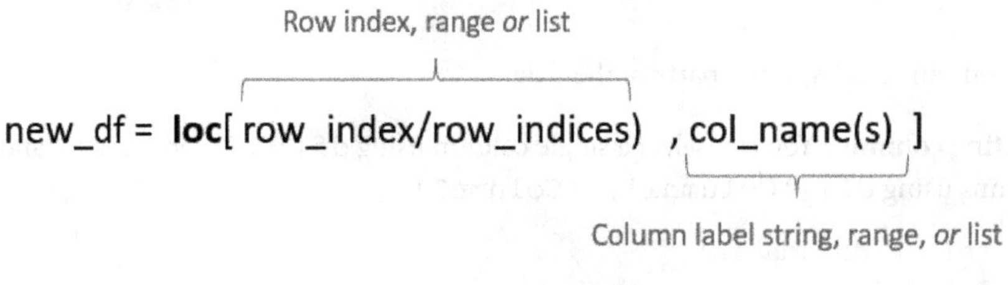

Figure 3.4: loc() syntax

Note

The first argument can be either a row index, range or list. Similarly, the second argument can be a label string, range, or list.

You can see the `iloc()` syntax here:

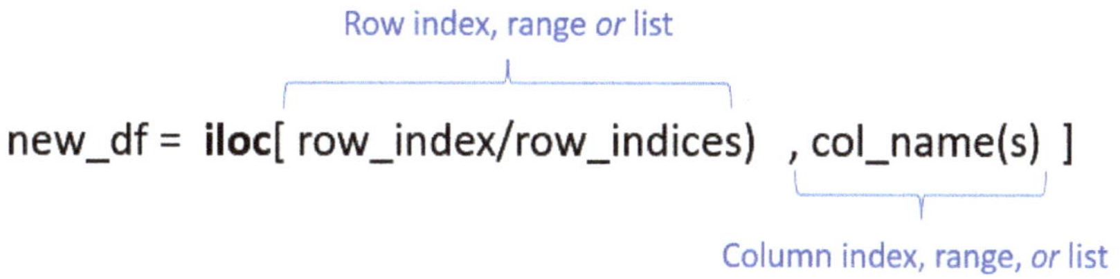

Row index, range *or* list

new_df = **iloc[** row_index/row_indices) , col_name(s) **]**

Column index, range, *or* list

Figure 3.5: iloc() syntax

Let's look at some examples using both of these. First let's create a dataset:

```
import pandas as pd

# Create a sample DataFrame
data = {
    'Name': ['John', 'Alice', 'Bob', 'Emily', 'Jack'],
    'Age': [25, 30, 35, 28, 32],
    'City': ['New York', 'London', 'Paris', 'Sydney', 'Tokyo'],
    'Salary': [50000, 60000, 70000, 55000, 80000]
}
```

Now, let's review how we can use `loc()` to select columns and/or rows:

```
df = pd.DataFrame(data)

# Select specific columns using loc()
selected_columns_loc = df.loc[:, ['Name', 'City']]
print("Selected columns using loc():")
print(selected_columns_loc)
print()
```

Next, we select the same information, using the `iloc()` method:

```
# Select specific columns using iloc()
selected_columns_iloc = df.iloc[:, [0, 2]]
print("Selected columns using iloc():")
print(selected_columns_iloc)
print()

# Select specific rows using loc()
selected_rows_loc = df.loc[1:3, :]
```

```
print("Selected rows using loc():")
print(selected_rows_loc)
print()

# Select specific rows using iloc()
selected_rows_iloc = df.iloc[2:4, :]
print("Selected rows using iloc():")
print(selected_rows_iloc)
print()

# Select a range of rows and specific columns using loc()
selected_range_loc = df.loc[1:3, ['Name', 'Age', 'Salary']]
print("Selected range of rows and specific columns using
loc():")
print(selected_range_loc)
print()

# Select a range of rows and specific columns using iloc()
selected_range_iloc = df.iloc[2:4, [0, 1, 3]]
print("Selected range of rows and specific columns using
iloc():")
print(selected_range_iloc)

)
```

Sorting data

Sorting in Python using the pandas library is a powerful technique that allows you to organize and analyze data efficiently. pandas provides various functions and methods to sort datasets based on one or multiple columns, thereby gaining insights from the data in a structured manner.

To perform alphanumeric sorting in Python using pandas, use `sort_values()` to specify the columns you want to sort by and the desired sorting order. Here is an example:

```
import pandas as pd

# Create a sample DataFrame
data = {
    'Name': ['John', 'Emma', 'Alex', 'Sarah'],
    'Age': [28, 32, 25, 30],
    'Salary': [5000, 7000, 4500, 6000]
}
df = pd.DataFrame(data)
```

```
# Sort the DataFrame by the 'Age' column in ascending order
sorted_df = df.sort_values('Age')

print(sorted_df)
```

Here is the output:

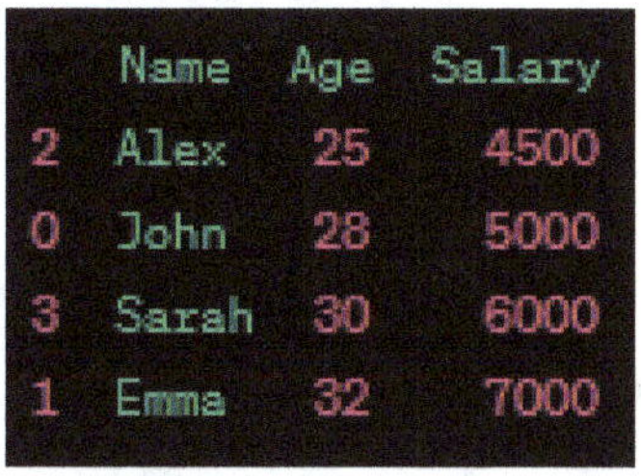

	Name	Age	Salary
2	Alex	25	4500
0	John	28	5000
3	Sarah	30	6000
1	Emma	32	7000

Figure 3.6: Sorting example 1

You can also sort by more than one column, as seen in the following example:

```
sorted_df = df.sort_values(["Age", "Salary"], ascending=[True, False])
```

The `ascending` parameter allows you to specify which columns should be sorted in ascending order. A value of `True` will ensure the respective column is sorted in ascending order; `False` will ensure that the column is instead in descending order.

This method also has another parameter called `na_position`. This method allows you to determine how `NA` values should be treated in the sorting process. For instance, setting this parameter to `first` means that `NA` values will appear at the top of the DataFrame. Here is an example:

```
sorted_df = df.sort_values('Age', na_position='first')
```

Here is the output:

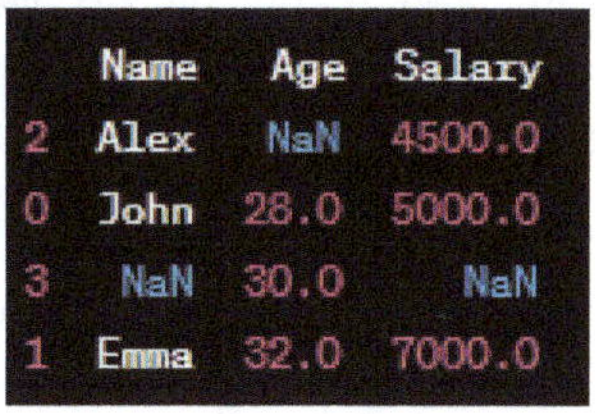

	Name	Age	Salary
2	Alex	NaN	4500.0
0	John	28.0	5000.0
3	NaN	30.0	NaN
1	Emma	32.0	7000.0

Figure 3.7: Sorting example 2

Merging data

The pandas library provides various facilities for efficiently combining dataframe objects. In particular, `merge` is a powerful function that allows us to perform database-style merging (or joining) operations (similar to `JOIN` operations in SQL).

Let's say you have two dataframes that share a common column key. A **key** is a column that's used to establish a relationship between two or more datasets. When joining data, the key serves as a common identifier or attribute that exists in both datasets, allowing for the combination of relevant information.

The process of joining data involves matching records from different datasets based on their key values. This enables the creation of a consolidated dataset that contains information from multiple sources.

In Python, the pandas library provides powerful tools for joining and merging DataFrames. The key(s) used for joining are specified through the `on` parameter, which accepts one or multiple column names. Here's how you can merge them:

```
import pandas as pd

df1 = pd.DataFrame({
    'key': ['A', 'B', 'C', 'D'],
    'value': np.random.randn(4)
})

df2 = pd.DataFrame({
    'key': ['B', 'D', 'D', 'E'],
    'value': np.random.randn(4)
})

merged = pd.merge(df1, df2, on='key')
```

The resulting `merged` DataFrame contains the `df1` and `df2` rows, where the key column matches, with the `df1` and `df2` columns concatenated. By default, `pd.merge()` performs an inner join, which means only the keys present in both dataframes are merged. In our example, we merge the two dataframes using one common key, but you can also merge on multiple keys. If a key doesn't exist in either dataframe, the corresponding row is excluded from the result.

But `merge()` allows other types of join operations, similar to SQL. Although the following options are not an exhaustive list, these are the ones you'll use most often:

- An **inner join** (the default functionality) is a join that returns only the matching records from both datasets based on the specified key(s). Non-matching records from either dataset are excluded from the result. The resulting dataset contains only the common records between the datasets, as seen in the following code example:

  ```
  merge(df1, df2, how='inner')
  ```

- An **outer join** (also known as a full outer join) is a join that returns both the matching and non-matching records from both datasets based on the specified key(s). Records from one table without a matching record in the other table will be filled with `null` or `NaN` values in the resulting dataset, as seen in the following code example:

```
merge(df1, df2, how='outer')
```

- A **left join** (also known as a full left join) returns all the records from the left (or first) dataset and the matching records from the right (or second) dataset. Non-matching records from the right dataset are filled with `null` or `NaN` values. The resulting dataset includes all records from the left dataset and the common records from the right dataset, as in the following code example:

```
merge(df1, df2, how='left')
```

- A **right join** is like a left join but returns all the records from the right (or second) dataset and the matching records from the left (or first) dataset. The resulting dataset includes all records from the right dataset and the common records from the left dataset, as seen in the following code example:

```
merge(df1, df2, how='right')
```

Aggregation with groupby()

Aggregation is a fundamental operation in data analysis that allows you to perform a summarization operation (e.g., sum, average, min, max, and so on) on a range of selected data by a specified grouping. The `groupby()` function in pandas provides a powerful way of performing aggregations. The concept of the `groupby()` operation can be compared with the concept of *Group By* in SQL and the *Split-Apply-Combine* strategy in R.

There are a variety of aggregation functions (e.g., sum, mean, median, and so on). However, most aggregation operations involve these three steps:

1. **Splitting** the data into groups based on some criteria. This involves selecting one or more categorical fields to group the data by.

2. **Applying** a function to each group. This is the function that dictates the kind of aggregation that you want to perform. Some examples include sum, **minimum (min)**, **maximum (max)**, **count**, and more.

3. **Combining** the results into a data structure.

Let's consider a dataframe:

```
import pandas as pd

data = {
    'Company': ['GOOG', 'GOOG', 'MSFT', 'MSFT', 'FB', 'FB'],
```

```
    'Person': ['Sam', 'Charlie', 'Amy', 'Vanessa', 'Carl', 'Sarah'],
    'Sales': [200, 120, 340, 124, 243, 350]
}


df = pd.DataFrame(data)
```

If we want to find the total sales of each company, we can use groupby() using the following syntax:

```
dataset.groupby('<Group(s)>')['<Aggregated_Col>'].agg_function()
```

Here, the Group(s) parameter represents the categorical field by which you want to group the aggregated result. The Aggregation_Col parameter represents the numeric field on which you want to perform the aggregation. Lastly, agg_function() represents the function that you want to use to perform the aggregation.

Let's apply this syntax to our example:

```
by_comp = df.groupby('Company')['Sales'].mean()
print(by_comp.head) # Outputs the first 5 rows of the result dataset
```

This code will create a groupby object, and then call the mean function on this result. It will then output the average sales for each company. In this case, the groupby() function splits the data into groups based on the 'Company' column. The mean() function is then applied to each of these groups independently, and the results are combined back into a new dataframe.

Here are a few more examples of how you can use groupby():

```
# To get the sum of sales for each company
 df.groupby('Company')['Sales'].sum()



# To get the standard deviation of sales for each company
 df.groupby('Company')['Sales'].std()



# To get more detailed information about each group
 df.groupby('Company')['Sales'].describe()
```

In addition to these, you can use any function with groupby() as long as that function can operate on a dataframe or Series. This includes both built-in pandas and numpy functions, as well as custom functions and lambda functions that you define yourself.

You can also apply multiple aggregations with more than one function using `agg()`. Here are some examples:

```python
import pandas as pd

# Create a sample DataFrame
data = {
    'Name': ['John', 'Alice', 'Bob', 'Emily', 'Jack'],
    'Age': [25, 30, 35, 28, 32],
    'Salary': [50000, 60000, 70000, 55000, 80000]
}

df = pd.DataFrame(data)

# Aggregate multiple columns with different functions
aggregations = {
    'Age': ['mean', 'min', 'max'],
    'Salary': ['sum', 'mean']
}

result = df.agg(aggregations)
print(result)
```

Here is the output:

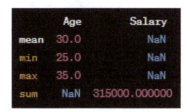

Figure 3.8: Aggregation output

Assessment

You are given the following dataframe with missing values:

```python
import pandas as pd
import numpy as np

df = pd.DataFrame({'A': [1, 2, np.nan],
                   'B': [5, np.nan, np.nan],
                   'C': [1, 2, 3]})
```

How would you fill the missing values in column `'A'` with the mean value of the non-missing values in the same column?

Answer

We use the `fillna` function of pandas dataframes, which allows us to replace NaN values with some value of our own:

```
df['A'].fillna(value=df['A'].mean(), inplace=True)
```

Here we are replacing NaN (missing) values in column `'A'` with the mean of non-NaN values in the same column.

Assessment

Given a dataframe `df` with a column called `'Company'` containing company names and a `'Sales'` column with their respective sales, write a code snippet that would filter out rows corresponding to the companies that have sales of more than 500.

Answer

Here, `df['Sales'] > 500` creates a Boolean Series where each element is true if the corresponding sales value is greater than `500`, and false otherwise:

```
df_filtered = df[df['Sales'] > 500]
```

This Series is used to index the original dataframe, resulting in a new dataframe with only the rows where the sales are more than 500.

Assessment

Suppose you have a dataframe `df` with the `'Name'`, `'Age'`, and `'Salary'` columns. How would you select the first three rows and the last two columns using the `iloc()` method?

Answer

To select the first three rows and the last two columns using the `iloc()` method, use the following code:

```
df.iloc[:3, -2:]
```

Assessment

Given a dataframe df with the 'Name', 'Age', 'Salary', and 'Country' columns, how would you select all rows where 'Age' is less than 40, and only select the 'Name' and 'Country' columns using both the loc() and iloc() methods?

Answer

To do this, you can use the following code:

```
df.loc[df['Age'] < 40, ['Name', 'Country']]
```

Or alternatively, use the following code:

```
df.iloc[df['Age'] < 40, [0, 3]]
```

Assessment

Suppose you have the following dataset:

```
import pandas as pd

data = {
    'OrderID': [1, 2, 3, 4, 5],
    'CustomerID': [101, 102, 103, 104, 105],
    'OrderDate': ['2022-01-01', '2022-02-15', '2022-03-10', '2022-04-
20', '2022-05-05'],
    'OrderTotal': [100, 150, 200, 75, 120]
}

df = pd.DataFrame(data)
```

How would you use the agg() function to calculate the total order amount for each customer?

Answer

First, we split the data into groups using groupby(); in this case, we need to group each customer. We then want to find the sum of the order totals, so we use the agg() method on the OrderTotal column. After that, we set our aggregation function to sum() since we want the total; this newly calculated column is given the name total_order_amount:

```
result = df.groupby('CustomerID')['OrderTotal'].agg(total_order_
amount=('sum'))
```

Assessment

You have two DataFrames, df1 and df2. Both DataFrames share a key column called "key". How would you merge these datasets?

Answer

The pd.merge() function is used to merge two DataFrames on a key:

```
merged_df = pd.merge(df1, df2, on='key')
```

By default, it performs an inner join, which means it will only include rows where the key is present in both df1 and df2. The resulting dataframe, merged_df, will include all columns from df1 and df2, but only rows where the key value is present in both.

Summary

In this chapter, we covered many Python programming fundamentals you would need for your technical interview. First, we covered Python variable data types and string operations, including string indexing. Afterward, we reviewed Python list comprehensions and control statements, including loops. Then we focused on some aspects of Python classes, indexing, merging, sorting, data aggregation, and handling missing data.

It is incredibly important to be proficient in the area of data wrangling and manipulation, which comprises a large part of data science interviews and assessments. Although it comprises a large part, data wrangling is tested proportional to its presence in data science jobs.

In the next chapter, we will move our focus from Python fundamentals to data visualization and storytelling.

References

- [1] *A Comparative Study of Data Cleaning Tools* by *Chen, Z., Oni, S., Hoban, S., & Jademi, O.,* from *International Journal of Data Warehousing and Mining (IJDWM) (2019).*

4

Visualizing Data and Data Storytelling

Data visualization is the process of creating images, charts, and other visual data. This is performed to reveal and understand underlying trends and patterns in the data. These skills are important in order for data scientists to tell compelling data stories. For example, a marketing analyst may examine online customer behavior to identify purchasing habit trends such as seasonal trends, product preferences, or demographic correlations. These patterns can be used to craft targeted marketing campaigns or develop personalized recommendations, enhancing customers. Alternatively, an analyst may analyze historical financial time series data to identify patterns in market trends, stock performance, or economic indicators. By recognizing patterns, they can make informed predictions about future market behavior, guide investment decisions, and develop risk management strategies.

In this chapter, you will delve into the world of data visualization and storytelling. Here, you will learn the key principles and techniques to choose the appropriate data visualization methods to effectively communicate insights and patterns hidden within complex datasets. The goal of this chapter is to equip you with the knowledge and skills necessary to create impactful and meaningful visual representations of data. By the end of the chapter, you will know some of the tools of the trade for data visualizations, including some software libraries, along with best practices for designing visually appealing and informative dashboards, reports, and **key performance indicators** (**KPIs**). Additionally, you will review coding techniques in Python that enable you to create charts and graphs programmatically. Lastly, we will introduce a framework for data storytelling, emphasizing the importance of narrative and context in presenting data-driven insights to various audiences.

Mastering these concepts is essential for you as a data scientist as it empowers you to effectively communicate your findings, influence decision-making, and inform business decisions across domains and industries.

In this chapter, we will cover the following topics:

- Understanding data visualization
- Surveying tools of the trade
- Developing dashboards, reports, and KPIs
- Developing charts and graphs
- Applying scenario-based storytelling

Understanding data visualization

As data scientists, we sometimes feel like we are explorers navigating the wild frontiers of massive datasets, hunting for insightful patterns and significant relationships. Yet, the real value of our journey lies in the capacity to translate these discoveries into stories that influence decisions, inspire action, and propel innovation. This is where the art of data visualization and storytelling comes into play.

Data visualization is a powerful tool beyond simply showcasing statistics or trends – it breathes life into data, transforming numbers and variables into visual narratives that capture attention, invoke emotion, and provoke thought. It is a translation process, converting the abstract language of data into an intuitive, visual dialect that people can understand and engage with. More than mere graphics, well-crafted data visualizations can tell compelling stories.

The power of visualization lies in its appropriateness to the data, the narrative, and the audience. Choosing the correct visualization is an important skill for data scientists – one that can significantly affect the comprehension, impact, and engagement of your data narrative. Let's take a look at different types of data visualizations.

Bar charts

A **bar chart** is a versatile visual that can display categorical data or discrete quantities. It compares different groups by representing them as rectangular bars with lengths proportional to the values they represent. A typical bar chart looks like this:

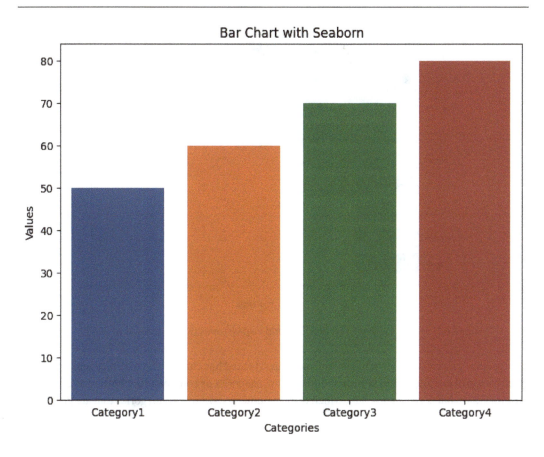

Figure 4.1: A bar chart

As well as being vertical, bar charts can be inverted horizontally to create a side-ways bar chart:

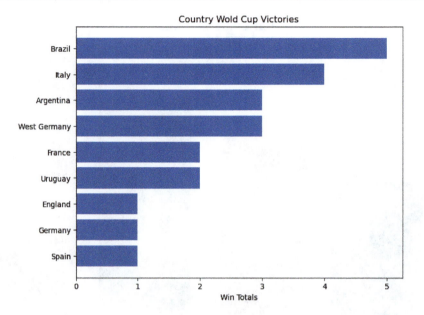

Figure 4.2: A sideways bar chart of FIFA World Cup victories by countries as of 2023

There are also numerous flavors of bar charts, including **stacked bar charts**, which display multiple bars stacked on top of one another to represent different subcategories or components within each category, or **grouped bar charts** (see *Figure 4.3*), which display numeric data across categories that are grouped.

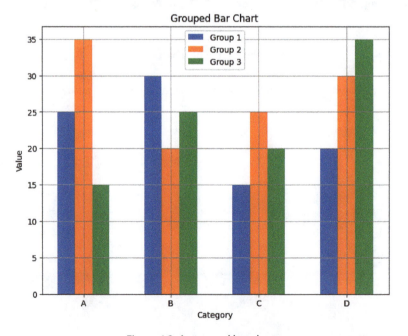

Figure 4.3: A grouped bar chart

When to use it: Bar charts are excellent for comparing quantities across categories, illustrating differences over time for a small number of groups, or presenting relative proportions.

Tips: Start the Y-axis at zero to avoid misrepresenting differences. Be sure to use appropriate scales to clearly illustrate differences in category quantities. Use horizontal bar charts when the category labels are long or if you have a large number of categories.

Line charts

A **line chart** represents quantitative data for one or more variables, making it ideal for showing the relationship between two quantitative variables (one for each axis), or displaying trends over time (where the X axis represents time). The plot is constructed by connecting data points with a line.

Figure 4.4 shows a typical line chart, where each axis represents numeric variables:

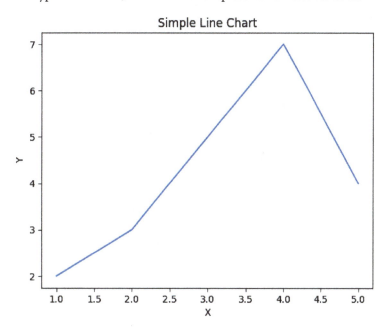

Figure 4.4: A line chart

As previously mentioned, line charts are sometimes used to create **time series plots**, which are a special type of line chart. Time series plots are simply line charts where a series of times (minutes, days, months, years, and so on) is the X-axis variable. While normal line charts are used to show the relationship between two numeric variables, time series plots specifically demonstrate the relationship between some numeric variable and time.

An example of a time series line chart is shown in *Figure 4.5*:

Figure 4.5: Time series line chart of stock price

When to use it: Use line charts to display trends, movements, or changes over time, or to compare the trends of different groups. Sometimes, line charts are used to assess whether two or more variables are correlated, or if they meet some data shape, such as exponential or logarithmic. For example, two variables may show a line chart that indicates exponential decay, which is useful when assessing numeric variable relationships.

Tips: Keep your chart simple; too many lines can make the chart hard to interpret. Use markers for each data point for added clarity.

Scatter plots

A **scatter plot** uses dots to represent values for two different variables, plotted on the X and Y axes. It allows for the observation of relationships or correlations, and is often the precursor to line charts if a pattern persists.

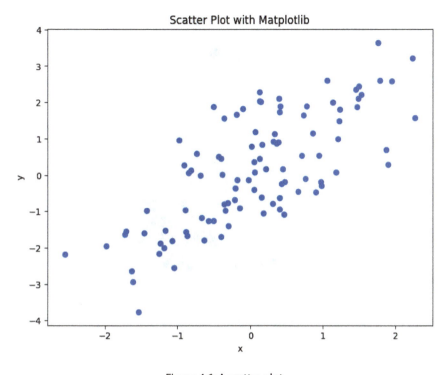

Figure 4.6: A scatter plot

When to use it: Scatter plots are perfect for showing relationships between two quantitative variables or displaying the distribution of data. They are handy when you want to highlight the correlation, or lack of any correlation, between two variables.

Tips: Use different colors or shapes to represent different categories (this may even uncover natural segments of data not previously known). Adding a trend line can help visualize the overall relationship.

Histograms

A **histogram** is a graphical representation of the distribution of data. It is used to display discrete numeric data, where the bins (or bars) represent ranges of data. It consists of a series of bars, where each bar represents a category or range of values, and the height of the bar represents the frequency or count of observations falling within that category. The bars in a histogram are typically placed adjacent to each other to emphasize the discrete nature of the variable. Histograms are useful for understanding the frequency and spread of values within different categories and identifying patterns or outliers in the data:

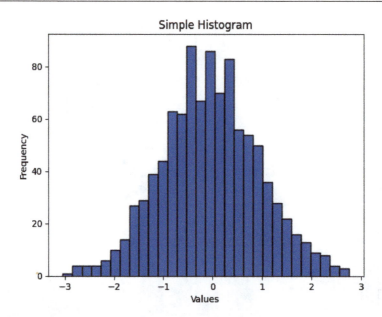

Figure 4.7: A histogram (approximately normal)

When to use it: Use histograms to show the distribution of (at least) one numeric variable and identify patterns such as **skewness**, **kurtosis**, or outliers.

Skewness is simply the degree of asymmetry. For example, a right-skewed histogram has a longer tail on the right side of its peak, whereas a left-skewed histogram has a longer tail on the left side of its peak. Alternatively, kurtosis is a measurement of deviation from a normal distribution. You'll learn more about distributions in the *Chapter 8*.

Tips: Be aware that bin size can greatly influence your histogram's shape and insights. Experiment with different sizes to find the one that best represents your data.

Density plots

Similarly, a **density plot** (also known as a **kernel density plot**) is another visualization method that's used to display the distribution of numeric data. Unlike histograms, density plots are used to represent the distribution of one or more continuous variables.

Alternatively, histograms are used to represent the distribution of discrete variables (which have a finite number of distinct values or categories). Thus, density plots provide a smooth estimate of the underlying **probability density function** (**PDF**) of the data. The plot displays the relative frequency of data points within different intervals along the variable's range, showing the concentration of data and areas of high or low density:

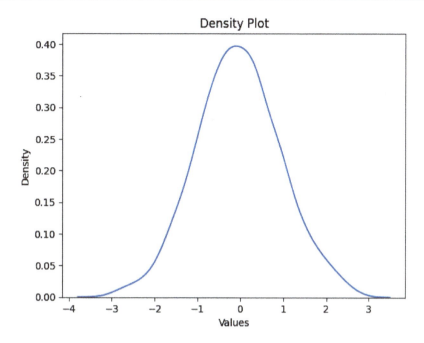

Figure 4.8: A density plot

When to use it: Use density plots to show the distribution of continuous variables and identify patterns such as skewness, kurtosis, or outliers. It is also useful while investigating the theoretical distribution of a variable.

We will learn more about common theoretical distributions in *Chapter 8*, but note that distributions derived from real-world empirical data are known as empirical distributions, which are in turn compared to theoretical distributions that determine data assumptions.

Tips: Similar to the histogram, you can create density plots with more than one variable to compare variable distributions and skewness.

Quantile-quantile plots (Q-Q plots)

A **Q-Q plot** is another plot that's used to assess a dataset's distribution, typically to compare it to some theoretical distribution (for example, normal distribution). It compares the quantiles of the empirical date from the dataset (along the *Y*-axis) against the quantiles of the expected theoretical distribution (along the *X*-axis). The diagonal line in this plot represents where the distributions would match exactly – the closer the scattered dots are to the line, the closer the dataset meets the theoretical distribution. In the case of the standard normal theoretical distribution, the expected quantiles would portray a mean of 0 and a standard deviation of 1:

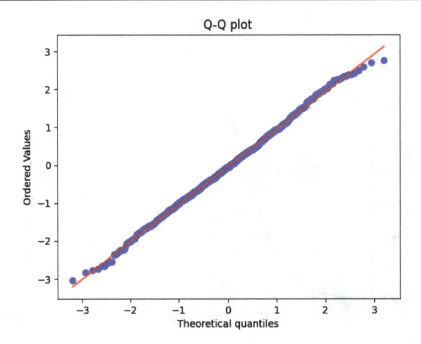

Figure 4.9: A Q-Q plot

When to use it: Q-Q plots are used to check underlying statistical assumptions, typically in preparation for statistical models that have distribution requirements, such as linear regression. The plot allows analysts to visually review the Q-Q plot to determine if the data meets pre-modeling requirements – that is, checking if the data fits some pre-determined theoretical distribution.

Q-Q plots are not as intuitive for visuals beyond pre-statistical analysis, so limit their use beyond your investigative needs. Non-analysts are much more likely to understand hisrograms and density plots.

Tips: When assessing a dataset's distribution, you may start with a histogram, which can give you some idea of the data's general distribution. If you are hoping for a normal distribution and the histogram shows that the data is skewed, there's not much point in using a Q-Q plot. However, if the histogram shows an approximately normal shape, you may use the Q-Q plot to give you a more accurate estimation of the distribution of data since it directly compares your data to the known quantiles of a theoretical distribution.

Box plots

A **box plot**, or box-and-whisker plot, provides a five-number visual summary of a dataset – the minimum, first quartile, median, third quartile, and maximum:

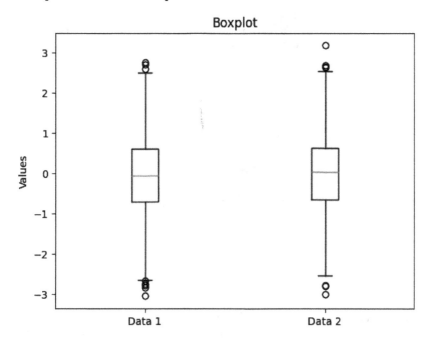

Figure 4.10: A box plot

When to use it: Box plots are great for comparing distributions between different groups or identifying outliers in your dataset. It requires at least one continuous numeric variable, but you can also plot numerous box plots by category. Box plots are better suited for summarizing the central tendency, spread, and identifying outliers, making them useful for comparisons between different variables or groups.

Tips: Pair box plots with other plots, such as a swarm plot, to show individual data points and give a more comprehensive view. **Swarm plots** are box plots overlaid with a scatter plot of the data.

Pie charts

A **pie chart** is a circular graph that represents proportions or percentages among categories, with each slice corresponding to a category, and all category proportions adding up to 100%. While pie charts are seemingly easy to interpret, most analytics professionals avoid them, since they have a way of tricking the human mind. We don't always correctly perceive the numeric proportions, making it difficult to make out categorical differences:

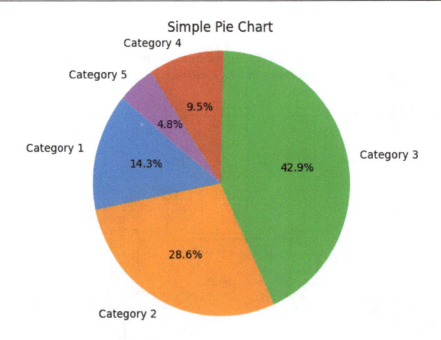

Figure 4.11: A pie chart

Additionally, it only takes a handful of categories before interpreting pie charts becomes extremely challenging. Furthermore, unlike bar charts, there's no natural ordering functionality.

When to use it: Traditionally, pie charts are suitable for displaying the proportions or a percentage of a whole regarding a small number of categories. However, it is advised to use them sparingly. Instead, consider using a bar chart or stacked bar chart.

Tips: Limit the slices to a manageable number (ideally under seven) to avoid overcomplicating the chart. Label slices with their actual values or percentages for clarity.

These are just a few examples in the vast world of data visualization. Remember, the goal is not to use the most complex visualization but the one that most effectively communicates your data story to the audience. As you gain experience, don't be afraid to experiment with less conventional visualization types, such as heatmaps, treemaps, or radial charts. Also, always be sure to keep your audience, the data, and your narrative in mind.

Assessment

You are given a dataset with sales data for a chain of grocery stores. The dataset includes sales figures by store location, product category, and time (monthly for the past 2 years). You're asked to analyze and present the monthly trend of total sales and also compare the sales of different product categories. What types of data visualizations would you choose for this task and why?

Answer

To present the monthly trend of total sales, a line chart would be the most appropriate visualization. Line charts are perfect for displaying trends over time and can easily showcase increases or decreases in total sales over the given 2-year period.

To compare the sales of different product categories, a stacked bar chart would be a good choice. Each bar could represent a month, and the segments within each bar would represent the sales of different product categories. This would allow the audience to visually compare the sales of different product categories and understand how they contribute to the total sales.

Assessment

You are working with a dataset containing responses to a customer satisfaction survey. The survey includes customer demographic information (age, gender, location, and so on) and responses to a question about satisfaction level on a scale from 1 to 5 (1 = very dissatisfied, 5 = very satisfied). What type of data visualization would you use to present an overview of the satisfaction level responses, and why?

Answer

A histogram or a bar chart would be appropriate to show an overview of the satisfaction level responses. A histogram would be a good choice because it shows the distribution of a single variable (in this case, satisfaction level). It can provide a visual representation of which satisfaction level was selected most and least often, as well as the general distribution of responses.

Alternatively, an ordinal bar chart could also work well, considering satisfaction levels are discrete and ordered categories. Each bar would represent a satisfaction level (from 1 to 5), and the length of the bars would show the count of responses for each level. This visualization would provide a clear view of customer sentiment, allowing for easy comparison between the different categories.

Both of these visualizations would help you quickly understand customer satisfaction level responses by visually representing the distribution and frequency of each response category.

Surveying tools of the trade

There is an array of visualization tools available that cater to a variety of needs, skill sets, and use cases. This section will discuss several popular data visualization tools, including Power BI, Tableau, R's Shiny, and Python libraries such as Matplotlib and Seaborn, providing guidance on when to use one over another. However, the goal here is to help give you more general knowledge to prepare you for your technical interview on understanding when to choose a particular tool.

Power BI

Power BI is a business intelligence tool developed by Microsoft. It offers interactive visualizations with an interface simple enough for end users to create reports and dashboards.

When to use it: Power BI is very effective when dealing with large quantities of complex data sources, which requires considerable data wrangling or modeling. It's an excellent choice for businesses seeking to create interactive, user-friendly dashboards or for integrating analytics into existing Microsoft-based systems.

Tableau

Tableau is a data visualization tool that's widely used for its intuitive ability to create complex, interactive visualizations, reports, and dashboards.

When to use it: Tableau shines when working with large and complex datasets, particularly when you need to create interactive dashboards or complex visual narratives. It's an excellent tool for organizations whose primary users are business analysts or executives who want to interact with the data but don't necessarily have extensive data modeling skills.

Shiny

Shiny is a package from RStudio that allows R and, as of late, Python users to build interactive web applications, bringing the power of R's statistical capabilities to visualization.

When to use it: Shiny is the tool of choice when your data work requires heavy statistical analysis, and you want to create web-based interactive visualizations. If you're already comfortable with R, Shiny allows you to leverage your existing skills while creating sophisticated applications.

ggplot2 (R)

ggplot2 is an R package known for creating elegant and aesthetically striking visualizations. It implements a unique *grammar of graphics* approach that allows for powerful plot customization and has a strong online community of users.

When to use it: ggplot2 is excellent when you're working with data in R and when you need to create complex, customized visualizations. Its strength lies in its flexibility and the consistency of its output.

Matplotlib (Python)

Matplotlib is a multi-platform data visualization library built on NumPy arrays for Python. It's powerful and flexible, capable of creating nearly any type of chart or graph.

When to use it: Matplotlib is excellent for creating simple to moderately complex static plots. It works well for customizing plots for publications or presentations or working with other Python libraries (such as NumPy or pandas).

Seaborn (Python)

When to use it: Seaborn is particularly useful when enhancing Matplotlib visuals. It's an excellent tool for exploratory data analysis and making statistical plots look more attractive.

Choosing the right visualization tool depends on the complexity of your data, the nature of your task, your team's technical skills, and your project's specific requirements. It's beneficial to familiarize yourself with several tools, so you can choose the most suitable one for each data visualization challenge you encounter.

Assessment

You are a data scientist at a multinational corporation using Microsoft-based infrastructure. Your manager has asked you to perform an in-depth analysis of a complex, large-scale dataset to derive insights into the company's operations and present your findings to both the technical team and non-technical stakeholders. You are comfortable with both R and Python. Considering these circumstances, which data visualization tools might best suit your task, and why?

Answer

Given the dataset's complexity and scale and the corporation's Microsoft-based infrastructure, Power BI would be an excellent choice for this task. Power BI is well-integrated with Microsoft's ecosystem, enabling smooth data import and export from various Microsoft sources. It's capable of handling large-scale datasets and producing interactive dashboards, which can be highly beneficial for presenting insights to non-technical stakeholders in an accessible, interactive manner.

However, considering that some of the audience is technical and you are comfortable with coding, utilizing Python's Matplotlib and Seaborn libraries or R's ggplot2 for exploratory data analysis and making customized, complex statistical graphics could be beneficial. These tools offer more control and customization for your plots and can handle the statistical nuances that might be required in the in-depth analysis. So, in essence, a combination of Power BI for interactive dashboard creation and either Python or R for more custom and intricate visualizations would be a well-rounded approach.

Developing dashboards, reports, and KPIs

In some technical interviews, you are given a take-home technical task to complete, and this might include data visualization. In the previous section, we touched on some common dashboarding tools a data scientist might use. In this section, we will delve deeper into some best practices for your dashboards, reports, and KPIs.

As a data scientist, you're not only tasked with uncovering insights from data but also communicating these insights effectively. This often involves creating dashboards, reports, and **KPIs**. While the aesthetics of your visuals are important, clarity, accuracy, and usability should always take precedence. The following are some best practices to help you create effective dashboards and reports:

- **Prioritize clarity and simplicity**: Avoid cluttered or overly complex visualizations. Keep your dashboards and reports simple and intuitive. Stick to one primary message per chart and limit the number of visualizations on a single page or screen. Remember, the goal of your visualization is to clarify, not confuse.

- **Use appropriate titles and labels**: Every chart or graph should have a clear, descriptive title that communicates its main point. Axis labels should be succinct yet descriptive. Including units of measurement where applicable is also essential. Legends should be easily identifiable and placed strategically so as not to interfere with the data.

- **Select the right chart type**: We discussed this earlier, but it is worth mentioning here also. The type of chart you use should align with the nature of your data and what you want to communicate. Bar charts and line graphs are generally more intuitive and versatile, while pie charts and scatter plots might require more context or explanation. Don't force a particular type of chart onto your data; instead, let the data guide your visualization choices.

- **Use consistent design elements**: Maintain consistency in color schemes, fonts, and styles across your dashboards and reports. This doesn't mean everything has to look the same, but there should be a cohesive, professional appearance to your work. Consistency reduces cognitive load and helps users focus on the content.

- **Implement interactivity**: Interactivity can greatly enhance the user experience by enabling users to focus on areas of interest, explore the data, and gain personalized insights. Filters, dropdowns, and hover-over effects are common interactive elements in dashboards. However, ensure interactivity doesn't compromise the clarity or performance of the dashboard.

- **Align visuals with KPIs**: KPIs should be front and center in your reports or dashboards. They should be visually distinct and easily understandable at a glance. Use simple but effective visual cues to indicate performance (such as colors or directional indicators).

- **Iterate and gather feedback**: In an interview setting, this might not be possible. Some interviewers like to interact with the interviewee as if they were colleagues. If this is the case, then don't consider your dashboard or report as a one-and-done task. Gather feedback from the interviewer as if they were the end user. Understanding how users interpret and interact with your visualizations can provide valuable insights for improvement.

Remember, data visualization is an art as much as it is a science. Strive for clarity and simplicity, but don't be afraid to experiment and innovate. The more you practice, the more intuitive and effective your data visualization skills will become.

Assessment

You've been asked to design a dashboard for a client who wants to monitor their website's traffic, user engagement, and sales performance. The client is not technically savvy, and the dashboard will be used by a diverse team within the organization, including sales, marketing, and product management.

The client's key metrics include the following:

- Daily and monthly unique website visits

- Average session duration

- Pages per session

- Sales conversion rate

- Top-performing products (by sales)

Considering the guidelines for creating effective dashboards, reports, and KPIs, outline your approach to designing this dashboard, including which visualizations you would use for each metric and how you would apply best practices to ensure the dashboard is effective and user-friendly.

Answer

The approach to this task should keep the end user in mind, making sure the dashboard is accessible, clear, and relevant to a broad audience within the organization:

- **Daily and monthly unique website visits**: Line charts would be ideal for tracking these metrics over time. They clearly show trends and fluctuations and would allow users to quickly understand the website's traffic patterns.

- **Average session duration**: Again, a line chart would be an effective visualization, providing an understanding of changes over time.

- **Pages per session**: A bar chart could be used here, possibly displaying average pages per session for each day or month.

- **Sales conversion rate**: A line chart tracking the sales conversion rate over time would be a clear way to display this important KPI.

- **Top-performing products**: A horizontal bar chart could effectively display top-performing products, making it easy for users to compare products.

To ensure the dashboard adheres to best practices, consider the following:

- Each chart should have a clear, descriptive title and labels, with units of measurement where necessary.

- A consistent color scheme and style should be used across all visualizations for a cohesive look. For example, all line charts could use the same color palette, distinguishing different lines with different shades or patterns.

- KPIs such as sales conversion rate and top-performing products should be highlighted and placed in prominent positions on the dashboard.

- To cater to the diverse audience and provide personalized insights, interactive features should be implemented. For instance, drop-down menus could be used to allow users to select specific time ranges or to filter products by categories.

- The design should be kept clean and uncluttered. If there are too many visualizations to fit comfortably on a single screen, tabs could be used to organize them into related groups.

Developing charts and graphs

While there are many tools for creating different data visuals, we will review a few basic visualizations, including bar charts, scatter plots, and histograms in Python. Two standard libraries for creating data visualizations in Python are Matplotlib and Seaborn.

In this section, we will discuss the different chart types and how to make them in Matplotlib and Seaborn.

Bar chart – Matplotlib

Matplotlib is a foundational library for visualizations in Python. Here's a basic example of how you might create a bar chart with Matplotlib:

```
import Matplotlib.pyplot as plt

# Categories and their associated values
categories = ['Category1', 'Category2', 'Category3', 'Category4']
values = [50, 60, 70, 80]

plt.figure(figsize=(8,6)) # Create a new figure with a specific size
(width, height)

plt.bar(categories, values) # Create a bar chart

# Labels for x-axis, y-axis and the plot
```

```
plt.xlabel('Categories')
plt.ylabel('Values')
plt.title('Bar Chart with Matplotlib')

plt.show() # Display the plot
```

Let's take a look at what the previous code block achieves:

- First, we import the Matplotlib library, specifically the `pyplot` module with the `matplotlibMatplotliby` convention. It's usually imported under the `plt` alias.

- Here, we're simply defining two lists: `categories` and `values`. These will be used for the X-axis (categorical data) and the Y-axis (quantitative data) of the bar chart, respectively.

- `plt.figure()` is a function that creates a new figure. The `figsize` parameter allows you to specify the width and height of the figure in inches.

- The `plt.bar()` function creates a bar chart. It takes two arguments: the X-values (our categories) and the Y-values (the corresponding values).

- The `plt.xlabel()`, `plt.ylabel()`, and `plt.title()` functions allow you to set labels for the X-axis, Y-axis, and the title of the plot, respectively. This step is crucial to make your plot self-explanatory.

- Finally, `plt.show()` is used to display the figure. It informs Python to display the figure and ensures that you can see it. This is necessary because Matplotlib is a graphical library and needs to interact with a graphical backend to display its figures.

Here is the result of the code block:

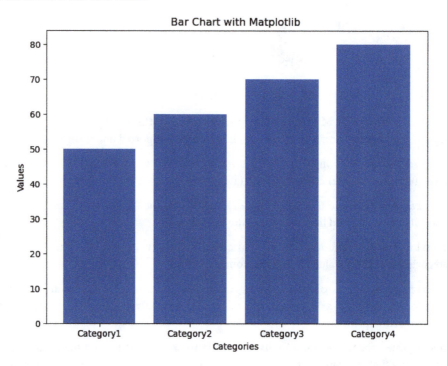

Figure 4.12: The output of the Matplotlib bar chart script

Bar chart – Seaborn

As mentioned previously, Seaborn is another Python library for data visualization built on top of Matplotlib. It allows us to layer in additional plotting features, such as adding colors or graphing themes, like so:

```
import Matplotlib.pyplot as plt
import seaborn as sns

# Categories and their associated values
categories = ['Category1', 'Category2', 'Category3', 'Category4']
values = [50, 60, 70, 80]

# Convert data to DataFrame
import pandas as pd
data = pd.DataFrame({"Categories": categories, "Values": values})
```

```
plt.figure(figsize=(8,6)) # Create a new figure with a specific size
(width, height)

sns.barplot(x="Categories", y="Values", data=data) # Create a bar
chart

# Labels for x-axis, y-axis and the plot
plt.xlabel('Categories')
plt.ylabel('Values')
plt.title('Bar Chart with Seaborn')

plt.show() # Display the plot
```

Let's review what the code block achieves:

- First, we import the Seaborn library, with `seaborn`. It is typically imported under the `sns` alias.

- Next, like the Matplotlib code, we define two lists, `categories` and `values`, that will hold the data we will plot. Then, we import the pandas library and create a DataFrame with our data. A DataFrame is a table-like data structure that Seaborn can use to create visualizations. Each key-value pair in the dictionary we pass to `pd.DataFrame` corresponds to a column in the DataFrame.

- The `plt.figure` function from Matplotlib is used to create a new figure. We specify the `figsize` parameter to set the width and height of the figure.

- The `sns.barplot` function is used to create a bar chart. We specify the columns of our DataFrame for the x and y parameters and pass our DataFrame to the `data` parameter. This tells Seaborn to create a bar chart with categories on the *X*-axis and values on the *Y*-axis.

- We use Matplotlib functions to add labels to our *X*-axis (`plt.xlabel`), *Y*-axis (`plt.ylabel`), and the title of the plot (`plt.title`).

- Finally, `plt.show` is used to display the plot. Seaborn relies on Matplotlib to display plots, and this function tells Matplotlib to render the following bar chart:

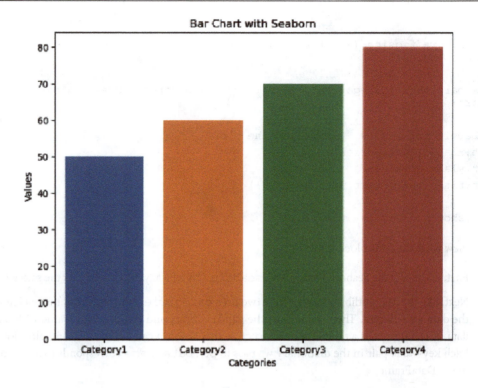

Figure 4.13: The output of the Seaborn bar chart script

Scatter plot – Matplotlib

Next, we will demonstrate how to create scatter plots with Matplotlib. Scatter plots are very useful for displaying relationships between two numeric variables along two different categories. It's often the first *sniff test* for investigating covariate relationships before applying more conclusive techniques such as regression analysis.

Let's take a look at how we might plot a scatter plot using the Matplotlib library:

```
import Matplotlib.pyplot as plt
import numpy as np

# Generate some example data
np.random.seed(0)
x = np.random.randn(100)
y = x + np.random.randn(100)
```

```
plt.figure(figsize=(8,6))

plt.scatter(x, y) # Create a scatter plot

# Labels for x-axis, y-axis and the plot
plt.xlabel('x')
plt.ylabel('y')
plt.title('Scatter Plot with Matplotlib')

plt.show()
```

Let's review the code:

- First, we import the Matplotlib library, specifically the `pyplot` module with the `matplotlibMatplotliby` convention. It's usually imported under the `plt` alias. Additionally, we import the NumPy module under the np alias to be used in creating a sample dataset later. NumPy is a library for the Python programming language that adds support for large, multi-dimensional arrays and matrices, along with a large collection of high-level mathematical functions to operate on these arrays.

- Additionally, we generate some random data for the scatter plot. `np.random.seed(0)` is used to keep the random numbers consistent between runs. `np.random.randn(100)` generates 100 random values from a normal distribution. In `y = x + np.random.randn(100)`, we generate the *Y*-values such that they have some relationship with the *X*-values (as they're based on x) but also have additional random noise.

- We create a new figure object where the plot will be drawn, with the size of the figure set to 8 units (width) by 6 units (height).

- To create the scatter plot, the `plt.scatter` function produces a scatter plot, with x and y being the data points that are plotted.

- The `plt.xlabel()`, `plt.ylabel()`, and `plt.title()` functions set labels for the *X*-axis, *Y*-axis, and the title of the plot, respectively.

- Again, we display the plot by calling the `plt.show()` function.

The following figure shows the result of the code block:

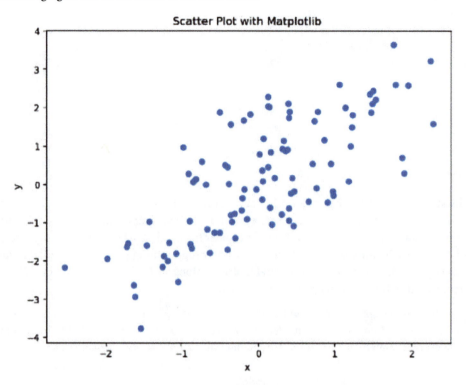

Figure 4.14: The output of the Matplotlib scatter plot script

Scatter plot – Seaborn

Now that we've explored scatter plots in Matplotlib, let's look at an example of using Seaborn:

```
import Matplotlib.pyplot as plt
import seaborn as sns
import pandas as pd
import numpy as np

# Generate some example data
np.random.seed(0)
x = np.random.randn(100)
y = x + np.random.randn(100)

# Convert data to DataFrame
```

```
data = pd.DataFrame({"x": x, "y": y})

plt.figure(figsize=(8,6))

sns.scatterplot(data=data, x="x", y="y") # Create a scatter plot

# Labels for x-axis, y-axis and the plot
plt.xlabel('x')
plt.ylabel('y')
plt.title('Scatter Plot with Seaborn')

plt.show()
```

Let's review the code:

- First, we import the necessary modules – that is, `seaborn`, `numopy`, and `pandas`.
- Next, the data is converted into a pandas DataFrame, a two-dimensional table data structure.
- Then, a new figure is created with a specified size.
- The `scatterplot()` function is called with our DataFrame passed to the data parameter and the column names for `x` and `y`.
- Finally, we add our labels to the plot and display it by calling the `plt.show()` function.
- This code will produce a scatter plot like the one produced by Matplotlib. However, Seaborn allows for more customization and complexity as you can map other variables to the size, hue, and style of the points, among other things.

Here is the result:

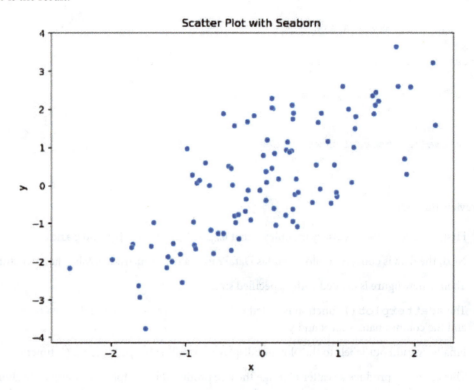

Figure 4.15: The output of the Seaborn scatter plot script

Histogram plot – Matplotlib

Moving on, here is an example of how we might create a standard histogram in Matplotlib:

```
import Matplotlib.pyplot as plt
import numpy as np

# Generate some example data
np.random.seed(0)
data = np.random.randn(1000)

plt.figure(figsize=(8,6))

plt.hist(data, bins=30) # Create a histogram

# Labels for x-axis, y-axis and the plot
plt.xlabel('Value')
```

```
plt.ylabel('Frequency')
plt.title('Histogram with Matplotlib')

plt.show()
```

Let's review the code:

- First, we import the Matplotlib and NumPy modules. Additionally, we generate random data with the NumPy library.

- Again, a figure is created with a specified size.

- A histogram is created with `plt.hist()`, with the data and number of bins as arguments.

- Finally, the *X*-axis, *Y*-axis, and the plot title are labeled and displayed.

- This code will produce a histogram with 30 bins.

Here is the result:

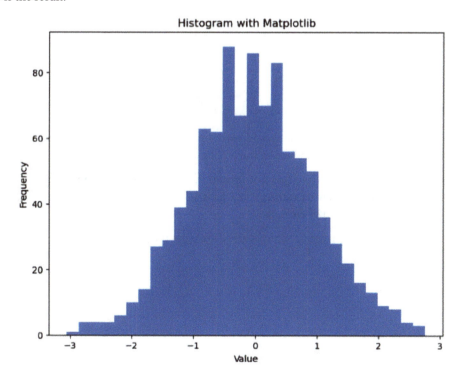

Figure 4.16: The output of the Matplotlib histogram plot script

Histogram plot – Seaborn

Next, let's see how Seaborn can add more character and visual appeal to our histogram plot:

```python
import Matplotlib.pyplot as plt
import seaborn as sns
import numpy as np

# Generate some example data
np.random.seed(0)
data = np.random.randn(1000)

plt.figure(figsize=(8,6))

sns.histplot(data, bins=30, kde=True) # Create a histogram

# Labels for x-axis, y-axis and the plot
plt.xlabel('Value')
plt.ylabel('Frequency')
plt.title('Histogram with Seaborn')

plt.show()
```

Let's explain the code:

- The first couple of lines of the code should look familiar as we import the necessary modules, generate some example data, and create the plot figure.

- The use of the `sns.histplot` function creates the histogram plot. In addition to plotting, with 30 bins, the **kernel density estimate** (**KDE**) is also plotted, which can help in visualizing the underlying distribution of the data.

- We complete the plot by labeling the axes and displaying the plot.

Here is the result:

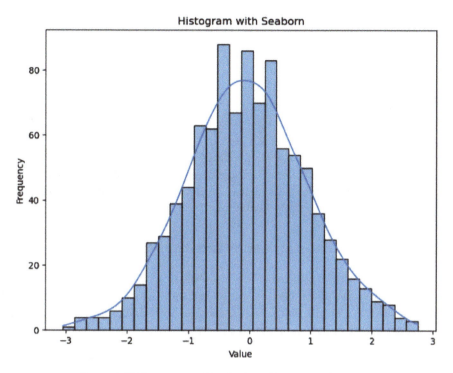

Figure 4.17: The output of the Seaborn histogram plot script

Assessment

You are given a dataset with a large number of data points and are tasked with visualizing the distribution of the values in the dataset. Which type of plot do you think is most suitable for this task in Matplotlib and Seaborn? How would you implement this in code?

Answer

A histogram would be most suitable for visualizing the distribution of values in a large dataset. Histograms provide a visual representation of data distribution by dividing the continuous data into bins and then plotting the number of data points that fall into each bin.

Here is an example of how this code might look:

```
import Matplotlib.pyplot as plt
import seaborn as sns

# Assuming data is your dataset
```

```
plt.figure(figsize=(8,6))

# Histogram with Matplotlib
plt.hist(data, bins=30)
plt.show()

# Histogram with Seaborn
sns.histplot(data, bins=30, kde=True)
plt.show()
```

In both the Matplotlib and Seaborn examples, the `bins` parameter determines the number of bins in your histogram (and can be adjusted based on your specific needs), and the `kde` parameter in the Seaborn example indicates whether or not to plot a Gaussian kernel density estimate (which can give you a smoother curve representative of the distribution).

Applying scenario-based storytelling

One of the most important aspects of a data scientist's role is to translate complex datasets into a narrative that people who aren't data scientists can understand. The ability to present your findings clearly and compellingly is a crucial skill for a data scientist. This section provides a framework for structuring your data story effectively:

- **Begin with your end**: Before crunching numbers, clarify your goal. What is the key message you want to communicate? What action do you want to take? A clear objective will guide your analysis, influence your choice of visualizations, and ensure your story resonates with your audience.

- **Know your audience**: Understanding your audience's needs, interests, and level of knowledge will help you present your data meaningfully. Tailor your story to fit your audience – the detail, complexity, and visualizations you use should vary depending on who you're speaking to.

- **Build your narrative**: Understanding your audience's needs, interests, and level of knowledge will help you present your data in a way that's meaningful to them. Again, here, we mention tailoring your story to fit your audience.

- **Use visuals wisely**: The human brain processes visuals much faster than text. Use this to your advantage by presenting your data visually. However, not all visuals are created equal. Your choice of visualization should simplify complex data, highlight the most important insights, and support your narrative. Keep your visuals clean and uncluttered, and avoid unnecessary decoration distracting from the data.

- **Let your data do the talking**: The best stories let the data speak for themselves. Use your narrative to guide your audience, but let the data provide the evidence. This makes your story more compelling and builds credibility and trust with your audience. Keep interpretations to a minimum to avoid speculating and making the data fit a preconceived story.

- **Engage and interact**: Make your data story interactive where possible. Allow your audience to explore the data for themselves, adjust the view, or filter the data. This makes your story more engaging and enables your audience to see the data from different perspectives.

- **Practice, review, and refine**: Like any form of communication, compelling data storytelling takes practice. Test your story on a trusted colleague or mentor. Ask for feedback and refine your story accordingly. Remember, the most effective data stories are not just accurate – they're also compelling:

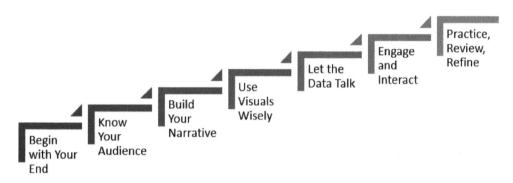

Figure 4.18: Scenario-based storytelling process

In summary, data storytelling is a powerful tool to enlighten your audience and drive action. But it's more than just presenting data and insights – it's about crafting a compelling narrative, choosing the right visuals, and letting your data speak for itself. As you grow in your data science career, remember that your ability to tell a compelling data story can be just as important as your technical skills.

Assessment

You've been asked to present an analysis of customer churn to the company's leadership. Considering the audience's high-level position and business-oriented mindset, how would you structure your presentation to ensure it is effective and engaging?

Answer

Given the audience's high-level position, it would be crucial to focus on the broader business implications of your analysis rather than on the technical details:

- Firstly, you'd want to clearly articulate the objective of your analysis (for example, to understand the reasons behind customer churn and propose strategies to reduce it) and ensure this aligns with the company's strategic goals.

- Secondly, you should build a clear and coherent narrative that guides your audience through your key findings and their implications, using relatable language and analogies where possible.

- Visuals should be used to effectively highlight the most critical insights – for example, a bar chart showing churn rates by customer segment or a line graph illustrating churn rates over time. These visuals should be clean, uncluttered, and easy to understand, focusing on the most critical data.

- Lastly, where possible, make your presentation interactive, perhaps by using a tool that allows your audience to explore the data further if they wish. Conclude with clear, actionable recommendations based on your data.

Assessment

In data storytelling, why is it important to "let your data do the talking," and how might you accomplish this when presenting your findings?

Answer

Letting your data do the talking means using data to provide evidence for your conclusions and to drive your narrative. It also means ensuring that your visualizations are accessible to a wide audience including non-technical personnel. In many cases, your visualization will be used by a wide variety of people from different backgrounds and role functionality. Thus, you must create visuals that require minimal explanation, that the intended insights from the plot are clear, and that you've considered accessibility (for example, colorblindness) in your development process. This approach ensures your story is grounded in facts, which lends credibility to your message and builds trust with your audience.

You can accomplish this by doing the following:

- Highlighting key data points and trends that support your message. This can be done visually, through graphs or charts, or narratively, by explicitly discussing these data points in your presentation.

- Keeping interpretations and conjectures to a minimum. While some interpretation of the data is usually necessary, it's important not to stray too far into speculation. Let the data drive the story rather than trying to make the data fit a preconceived narrative.

- Presenting raw numbers or statistics where appropriate. While visuals are often more engaging, sometimes, the most effective way to let your data speak is to present the numbers themselves, especially when those numbers are particularly impactful.

- Using direct quotes or anecdotes from qualitative data to emphasize or illustrate a point. This can make the data more relatable and personal, adding another dimension to your story.

Summary

In the first half of this chapter, we established the critical role of data visualization and storytelling in the field of data science. Beginning with an overview of why data visualization is crucial, we delved into a framework for choosing the right visualization based on data types and the goal of communication. We explored a variety of data visualization types, such as bar charts, pie charts, histograms, scatter plots, and box plots, discussing their use cases, creation processes, and tips for enhancing their storytelling power. Additionally, we analyzed various visualization tools, including Power BI, Tableau, R's Shiny, Python's Matplotlib, and Seaborn, providing insights into their advantages, limitations, and ideal use cases.

The latter part of this chapter focused on the practical aspects of data visualization and storytelling. We covered the best practices for creating effective dashboards, reports, and KPIs, emphasizing clean, uncluttered visuals, appropriate titles, readable axes, and interactivity. Hands-on implementation of different plots using Python's Matplotlib and Seaborn was extensively discussed, with explanations and code examples for creating bar charts, scatter plots, and histogram plots.

The final section emphasized the crucial role of storytelling, providing a clear framework for structuring a compelling data story. Throughout, you were equipped with assessment questions to reinforce your understanding, preparing you for job interviews and practical applications in your early career as a data scientist.

In the next chapter, we will look at preparing for the SQL-based questions of the technical interview.

5

Querying Databases with SQL

In this chapter, you'll learn the essential aspects of databases, starting with a broad overview, then diving deep into the fundamental language of SQL, exploring crucial concepts such as subqueries, JOIN, CASE WHEN, window functions, aggregations, and how to tackle complex queries.

Our goal is to provide you with the knowledge and tools necessary to tackle any database-related question in a technical interview effectively. This is crucial for those preparing for a technical interview because understanding databases is a foundational skill for data scientists; equipped with the knowledge shared in these chapters, you'll be ready to face any database question confidently and proficiently.

So, in this chapter, we will cover the following topics:

- Introducing relational databases
- Mastering SQL basics
- Aggregating with GROUP BY and HAVING
- Creating fields with CASE WHEN
- Analyzing subqueries and CTEs
- Merging tables with joins
- Calculating window functions
- Approaching complex queries

Introducing relational databases

A database is a critical component in data-driven businesses and organizations, and data scientists need to understand its structure, functions, and underlying language. This section aims to introduce you to relational databases, focusing on the common language of SQL.

A **relational database** (also known as a SQL database) is a type of database that organizes data into tables, where each table has rows and columns. Each table represents a specific entity type, such as **Customers** or **Products**. Much like DataFrames, each row represents a unique record (or records), and each column represents a field (or attribute) of the data. This relational model introduced a standard way to represent and query data independent of any specific application. You can see an example of a relational database in *Figure 5.1*:

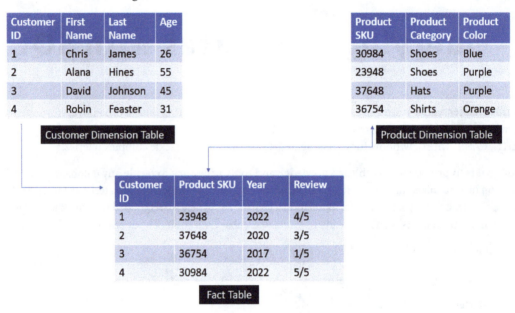

Figure 5.1: Relational data model example – star schema

What makes relational databases so powerful is their ability to establish efficient and useful relationships between multiple datasets so that they can be *joined* to create unique views and insights while ensuring data integrity.

To fully understand how relational databases work, let's examine some key concepts:

- **Primary key**: A primary key is a unique identifier for each record in a table. It serves as a means to uniquely identify and distinguish individual rows within a table by joining them to a like-primary key (called a *foreign key*) in another table. We will discuss joins later in this chapter.

- **Schema**: Schemas are standard data model structures and logic used in SQL databases. There are a handful of standard schemas, but there are a few that you will see 99% of the time:

  - **Star schema**: This schema consists of **fact tables** representing business events and **dimension tables** representing various attributes related to the facts. The fact table resides in the logical *center* of the data model and is connected to one or more dimension table(s) through primary and foreign key relationships. *Figure 5.1* demonstrates a star schema data model with one fact table and two dimension tables. The fact tables are connected to the dimension tables by the **Customer ID** and **Product SKU** keys.

- **Snowflake schema**: This schema is an extension of the star schema and is used to normalize dimension tables further. In a snowflake schema, dimension tables are divided into multiple levels, creating a more complex network of relationships. *Figure 5.2* demonstrates the structure of a snowflake schema using the same data as *Figure 5.1*, with some added detail. Instead of one fact table and two dimension tables, the snowflake schema expands on the dimension tables by giving them related dimension tables:

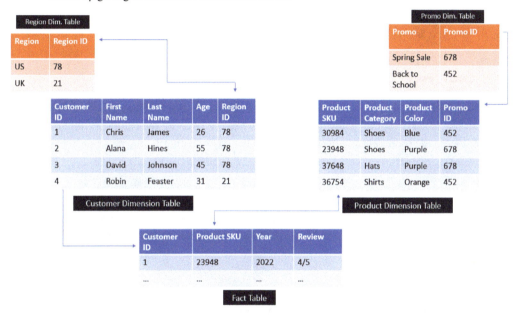

Figure 5.2: Relational data model example – snowflake schema

SQL is an indispensable tool for data scientists – it is used to query and manipulate the data stored in the database, plus it allows you to retrieve specific data, group it, sort it, and join different tables, all of which are key tasks in data analysis.

> **Note**
>
> While undergoing an SQL or database interview, be sure to ask the interviewer to specify the version of SQL with which you will be tested. It may also be worth reviewing the preferred SQL version so that you are optimally prepared.

Mastering SQL basics

As a data scientist, mastering the basics of SQL is crucial. Luckily for you, the basics are pretty easy to grasp, even for non-technical learners. This is because, at this stage, SQL generally reads like English sentences. To get you started, this section focuses on three fundamental components of SQL: the SELECT, WHERE, and ORDER BY statements.

The SELECT statement

The **SELECT** statement is the foundation of any SQL query and is used to retrieve data from a database. The general syntax is as follows:

```
SELECT column1, column2, ..., columnN
FROM table_name;
```

The syntax lists the different columns you want to return, separated by a comma. Since databases hold numerous tables, the query code specifies which table to select the columns using the FROM statement. Lastly, the semi-colon (;) is used to mark the end of a query.

> **Note**
>
> It is standard to create a new line of the query for each main clause (which is capitalized). Here, we started a new line once we began our FROM clause. Although this is not a hard rule, this structure is fairly standard, and it is advised that you follow it when needed to keep your code legible and organized.

Consider an example where we have a table named employees with the first_name, last_name, and salary columns. We can retrieve all the first names and last names with the following SQL query:

```
SELECT first_name, last_name
FROM employees;
```

It may get daunting listing columns, especially if you want to list more than a few. SQL provides a useful operator called the **wildcard** to return all the columns of a query's output. To use the wildcard, we must use *, like so:

```
SELECT *
FROM employees;
```

If your dataset has duplicates and you only want to return distinct values, use **DISTINCT** with SELECT. This clause is also an excellent method to display unique values in a column. For example, the following query shows all distinct breeds of dogs in the given table:

```
SELECT DISTINCT breeds
FROM dogs;
```

The WHERE clause

While the SELECT statement allows us to specify which columns we want to retrieve, the **WHERE** clause lets us define conditions to filter the rows being selected. The general syntax is as follows:

```
SELECT column1, column2, ..., columnN
FROM table_name
WHERE condition;
```

condition can involve various logical and comparison operators:

Operator	Meaning	Example	Explanation
=	Equal to	SELECT * FROM Table WHERE Name= 'Malik';	Returns rows where the Name column equals "Malik"
<>	Not equal to	SELECT * FROM Table WHERE NAME <> Malik;	Returns rows where the Name column does not equal "Malik"
<	Less than	SELECT * FROM Table WHERE Salary < '100000';	Returns rows where the Salary column is less than 100,000
>	Greater than	SELECT * FROM Table WHERE Salary > '50000';	Returns rows where the Salary column is less than 100,000
<=	Less than or equal to	SELECT * FROM Table WHERE Salary <= '100000';	Returns rows where the Salary column is less than or equal to 100,000
>=	Greater than or equal to	SELECT * FROM Table WHERE Salary >= '50000';	Returns rows where the Salary column is greater than or equal to 50,000
BETWEEN ... AND...	Between ... and ...	SELECT * FROM Table WHERE Salary BETWEEN 50000 AND 100000;	Returns rows where the Salary column is between the values 50,000 and 100,000
IN	Within (a list)	SELECT * FROM Table WHERE City IN ('Columbus','Chicago','Indianapolis');	Returns rows where the City column is equal to "Columbus" or "Chicago" or "Indianapolis".
OR	Returns rows which satisfy any clauses of the OR statement	SELECT * FROM Table WHERE Salary < '100000' OR City = 'Detroit';	Returns rows where the Salary column is less than 100,000 OR the City column is equal to Detroit (both do not need to be true).
AND	Returns rows which satisfy all clauses of the AND statement	SELECT * FROM Table WHERE Salary < '100000' AND City = 'Detroit';	Returns rows where the Salary column is less than 100,000 AND the City column is equal to Detroit (both must be true).

Figure 5.3: Common logical and comparison operators

The ORDER BY clause

Once we have selected the required data, we often want to order the results in a specific way. This is where the **ORDER BY** clause comes in – it sorts the result set by one or more columns. The general syntax is as follows:

```
SELECT column1, column2, ..., columnN
FROM table_name
ORDER BY column1 [ASC|DESC], column2 [ASC|DESC], ... columnN
[ASC|DESC];
```

The ORDER BY clause, by default, sorts the results in ascending order (ASC). If you want to sort the results in descending order, you can use the DESC keyword.

As you can see in the provided syntax, you can also order using multiple columns by separating the different columns with a comma. The first column will be ordered first, followed by the second column, and so on.

For example, to retrieve all employees and order them by salary in descending order and then age in descending order, the SQL query would be as follows:

```
SELECT *
FROM employees
ORDER BY salary DESC, age DESC;
```

Here is an example output for this query (given the employee_ID, first_name, last_name, salary, and age fields):

```
4 | Sophia | Davis | 6500 | 32
2 | Emily | Johnson | 6000 | 35
5 | Daniel | Jones | 6000 | 31
3 | Michael | Williams | 5500 | 28
1 | John | Smith | 5000 | 30
```

Notice that the data is ordered by salary first, then by age.

Assessment

Given a table named Products with ProductID, ProductName, Category, and Price columns, write a SQL statement to select all the products in the 'Electronics' category where the price is greater than 100. The results should be ordered by Price, in descending order.

Answer

Here's the answer:

```
SELECT *
FROM Products
WHERE Category = 'Electronics' AND Price > 100
ORDER BY Price DESC;
```

The SELECT * statement selects all columns in the Products table. The WHERE clause filters products that are in the 'Electronics' category and have a price greater than 100. Then, the ORDER BY clause orders the output by Price in descending order (from highest to lowest).

Assessment

Consider a table named `Orders` with `OrderID`, `CustomerID`, `ProductID`, and `Quantity` columns. Write a SQL statement to select `ProductID` values that have been ordered in a quantity greater than 5. The result should be ordered by `ProductID` in ascending order.

Answer

Here's the answer:

```
SELECT ProductID
FROM Orders
WHERE Quantity > 5
ORDER BY ProductID ASC;
```

The `SELECT` statement selects the `ProductID` column from the `Orders` table. The `WHERE` clause filters orders where the `Quantity` ordered is greater than 5. Then, the `ORDER BY` clause orders the output by `ProductID` in ascending order.

Aggregating data with GROUP BY and HAVING

Aggregation is a concept with which you should already be familiar thanks to the discussion of Python using pandas in *Chapter 3*. Just like in Python, aggregation in SQL is about summarizing or grouping data in a way that makes it more useful, understandable, and manageable. `GROUP BY` and `HAVING` are two crucial components in SQL that help accomplish this.

The GROUP BY statement

Much like how grouping is performed in Python using pandas, the **GROUP BY** statement in SQL is used with aggregate functions (such as `COUNT`, `SUM`, `AVG`, `MAX`, and `MIN`) to group the result set by one or more columns. Thus, using `GROUP BY` should be familiar to you! The syntax is as follows:

```
SELECT column1, column2, columnN aggregate_function(columnX)
FROM table
GROUP BY columns(s);
```

Aggregate values are best managed by using aliases. An **alias** is simply a nickname for a calculated or aggregated field or temporary table. Simply use the term `AS`, like so:

```
SELECT column1, aggregate_function(column2)AS alias
```

For example, let's say we have a table called `employees` with the `employee_id`, `first_name`, `last_name`, `salary`, and `department_id` columns. If we want to find out the total salary paid out by each department, we could write the following:

```
SELECT department_id, SUM(salary) as total_salary
FROM employees
GROUP BY department_id;
```

This query will return a list of department IDs, along with the total salary for each department. We assigned the `total_salary` alias to the sum of salaries.

> **Note:**
> Technically, you do not have to use the `AS` keyword to create an alias. You can simply provide the alias name immediately, like so: `SELECT column1, agg_function(column2) alias FROM table;`

Single-valued grouping rule

There is a little rule when it comes to using `GROUP BY` that will save you from a frustrating mistake if followed. The **single-valued grouping rule** dictates that any field included in the `SELECT` clause that is not part of an aggregate function should either be included in the `GROUP BY` clause or be part of a unique constraint in the table. This ensures that each column in the `SELECT` clause represents a single value for each group defined by the `GROUP BY` clause. Here is an example:

```
SELECT DepartmentID, DepartmentManager, COUNT(EmployeeID) AS
EmployeeCount
FROM Employees
GROUP BY DepartmentID, DepartmentName;
```

In this example, we must group by both `DepartmentID` and `DepartmentManagers` to return the number of employees for each unique combination of department ID and department manager.

Here is an example output:

DepartmentID	DepartmentManagers	EmployeeCount
1	Anya	8
1	Lola	12
2	Dustin	24
3	Cody	15

Figure 5.4: The single-valued grouping rule applied

However, there are exceptions to this rule. In some cases, if a field in the `SELECT` clause is functionally dependent on a column that is already part of the `GROUP BY` clause, it does not need to be included explicitly. Consider this example, where we want to return the max order amount for each customer, and their corresponding order dates:

```
SELECT CustomerID, OrderDate, MAX(TotalAmount) AS MaxOrderAmount FROM
Orders
GROUP BY CustomerID;
```

Here is an example output:

CustomerID	OrderDate	MaxOrderAmount
101	2023-01-02	100
102	2023-01-03	200
103	2023-01-03	200

Figure 5.5: Single-valued grouping not applied

This example does not follow the single-valued grouping rule because we want the max order amount for each `CustomerID`, but not for each unique order date. Thus, the `MAX` function will calculate the maximum total order amount for each unique customer, but not for each customer's unique order date. The result is the max order amount for each unique customer and that order's corresponding order date.

> **Note**
>
> Not all database systems handle this exception consistently, so it's generally recommended to follow the single-valued grouping rule for portability and clarity.

The HAVING clause

The **HAVING** clause was added to SQL to filter the results of the `GROUP BY` clause since `WHERE` does not work with aggregated results. The syntax for the `HAVING` clause is as follows:

```
SELECT column1, aggregate_function(column2)
FROM table
GROUP BY column1
HAVING aggregated_condition;
```

Suppose we want to find out which departments have a total salary payout greater than 50,000. We would enter the following code:

```
SELECT department_id, SUM(salary) as total_salary
FROM employees
GROUP BY department_id
HAVING SUM(salary) > 50000;
```

In this query, the HAVING clause filters out the groups (in this case, departments) for which the total salary is not greater than 50,000.

GROUP BY and HAVING are fundamental components of SQL, especially when working with large datasets.

> **Note**
>
> The HAVING clause is similar to the WHERE clause – so similar that novice SQL learners are confused regarding when to use one over the other. So, let's make the distinction between the two now.
>
> The WHERE clause is used in a SELECT statement to filter rows based on specified conditions *before* the data is grouped or aggregated. It operates on individual rows and filters them based on the given conditions.
>
> The HAVING clause is used in combination with the GROUP BY clause in a SELECT statement to filter rows based on specified conditions *after* the data is grouped and aggregated. It operates on the result of the grouping operation and filters the aggregated data.

Assessment

Consider these two tables:

- Employees, with columns for EmployeeId, FirstName, LastName, and DepartmentId
- Departments, with columns for DepartmentId and DepartmentName

Write a SQL query to find out departments that have more than five employees with a salary greater than 65,000.

Answer

Here's the answer:

```
SELECT d.DepartmentName
FROM Employees e
INNER JOIN Departments d ON e.DepartmentId = d.DepartmentId
```

```
WHERE e.Salary > 65000
GROUP BY d.DepartmentName
HAVING COUNT(e.EmployeeId) > 5;
```

In this query, INNER JOIN is used to combine rows from Employees and Departments where DepartmentId matches in both tables. The WHERE clause filters employees with a salary greater than 65,000. The GROUP BY clause groups the remaining data by DepartmentName. The HAVING clause is then used to filter these groups to include only those with more than five employees.

Creating fields with CASE WHEN

The **CASE WHEN** statement is a straightforward technique for creating new fields using conditional logic. It allows you to specify multiple conditions and define actions or outcomes for each condition. The CASE WHEN statement is often used to transform data, create calculated columns, or perform conditional aggregations. The syntax of the CASE WHEN statement is as follows:

```
CASE WHEN
condition1 THEN result1
WHEN condition2 THEN result2
WHEN conditionN THEN resultN
ELSE else_result
END As alias;
```

Here is an example where we create a new field that will detail if a student passed or failed, based on their scores:

```
SELECT student_id, student_name, exam_score,
CASE WHEN exam_score >= 60 THEN 'Pass'
ELSE 'Fail'
END AS result
FROM students;
```

This query creates a new field called result, and populates it with "Pass" when the student scored at least 60 on their exa; otherwise, it populates it "Fail". The results are returned with the student's name and ID.

Analyzing subqueries and CTEs

SQL **subqueries**, also known as nested queries or inner queries, are queries that are embedded within the context of another SQL query. They are powerful tools for performing complex data manipulations that require one or more *intermediary* steps – that is, they are used for performing data manipulation operations that require multiple steps or depend on the result of an intermediary query.

That might sound complex and, indeed, subqueries can easily get very complicated. But once you know the rules of engagement, you'll soon see that they're very doable. Before implementing a subquery, ask yourself the following:

- Where am I starting?

- Where am I going?

If you can answer these two questions, you've won half the battle. The other half is determining what steps need to take place to get from point A (existing data) to point B (desired data). In this section, we will learn how to navigate multi-step queries with subqueries.

We'll begin our journey by examining the different types of subqueries:

- **SELECT subqueries**: Where the subquery is located in the SELECT clause

- **FROM subqueries**: Where the subquery is located in the FROM clause

- **WHERE subqueries**: Where the subquery is located in the WHERE clause

- **HAVING subqueries**: Where the subquery is located in the HAVING clause

In the following subsections, we will review the use of subqueries in the SELECT, WHERE, FROM, and HAVING clauses.

Subqueries in the SELECT clause

Subqueries in the SELECT clause are the easiest to grasp because it feels similar to how we used SELECT in the past. Historically, we simply use SELECT to return a specific column or an aggregate of a column (for example, SUM). Selecting a subquery is useful when we want to return something that doesn't currently exist, hence the need for at least one additional, intermediary step. Consider the following questions while building a query:

- Is the desired output selectable? (In other words, is it an existing field?)

- Is the desired output a single-step calculation (for example a SUM (column) or CASE WHEN use case)?

If the answer to both of the previous questions is no, it may be a job for a SELECT clause subquery!

Here's how it works:

```
SELECT column1, column2, columnN,
(SELECT agg_function(column) FROM table WHERE condition)
FROM table
```

This code returns the specified columns. One of the specified columns is a subquery, which uses an aggregation function to summarize a column in the original table.

Here is a more concrete example:

```
SELECT CustomerID, SUM(TotalAmount) AS TotalSales, (
SELECT COUNT(*)
FROM Orders
WHERE CustomerID = O.CustomerID AND TotalAmount > 1000) AS
HighTotalAmountOrderCount
FROM Orders O
GROUP BY CustomerID;
```

In this example, we can see the following:

- The subquery within the SELECT clause calculates the count of orders where TotalAmount is greater than 1,000 for each customer:

  - This count is specific to each customer as it correlates with the outer query using the CustomerID column

  - The result of the subquery is aliased as HighTotalAmountOrderCount and displayed as a single column in the result set

- The outer query retrieves CustomerID and the aggregated sum of TotalAmount as TotalSales for each customer:

  - GROUP BY groups the results by CustomerID

Subqueries in the FROM clause

Subqueries in the FROM clause create a temporary table that can be used for the main query. This allows the programmer to simplify the process by breaking the problem into smaller, more manageable parts.

To master subqueries in the FROM clause, be sure to identify your current dataset and the desired output. From there, it's a process of molding your current data into the desired data in steps. Here is an example:

```
SELECT employee, total_sales
FROM (SELECT first_name || ' ' || last_name as employee, SUM(sales) as
total_sales
      FROM sales
      GROUP BY employee) as sales_summary
WHERE total_sales > 100000;
```

In this example, the subquery creates a temporary table aliased as `sales_summary`, which does the following:

- *Concatenates* each employee's first and last name (separated by a space). This concatenation is aliased as `employee`.

- *Calculates* the total sales for each employee.

- *Groups* the `total_sales` by employee.

So, even without knowledge of the `sales` table values, we know that the structure of the output will look something like this:

employee	total_sales
...	...
...	...
...	...

Figure 5.6: Intermediary results of the subquery

The outer query then selects `employee` and `total_sales` from the `sales_summary` temporary table. These results are filtered to all employees who made total sales greater than $100,000.

Subqueries in the WHERE clause

Subqueries in the WHERE clause are used to filter rows based on conditions detailed in a subquery. This method is useful when you don't already have access to the condition on which you want to filter your query.

Consider our previous example, where we performed the following basic query:

```
SELECT *
FROM Table
WHERE Salary < '100000';
```

In this example, we filtered the results from our table to rows, where the `Salary` field is less than `100000`. This is a single scalar value, which is available to us via the hardcoded `100000` value. But what if the condition isn't readily available? What if we needed to derive the condition since it doesn't already exist, or perhaps this condition isn't scalar? Perhaps the condition is dynamic? This is the power of subqueries in the WHERE clause.

> **Note**
>
> In the context of subqueries, the inner query is the subquery, and the **outer query** is the query that's querying the subquery. What a mouthful! Keep in mind that the innermost query is always evaluated first.

WHERE subqueries are most commonly used with scalar values or non-scalar values as the condition. In this context, a **scalar value** is the result of a subquery that yields one single value. Alternatively, a **non-scalar value** is a subquery that returns a 0 (False) or 1 (True) to return true values.

> **Tip**
>
> Always read SQL queries from the inside out by reading the innermost query first and working your way out.

Scalar example

Let's take a look at a scalar example. Suppose that we have a table called employees with employee_id, first_name, last_name, salary, and department_id columns. If we want to find all employees who earn more than the average salary, we can use a subquery:

```
SELECT first_name, last_name, salary
FROM employees
WHERE salary > (SELECT AVG(salary) FROM employees);
```

The subquery (SELECT AVG(salary) FROM employees) calculates the average salary of all employees, which is a scalar value. The outer query filters each row by the condition that it is greater than the average salary of all employees. The results are the first_name, last_name, and salary values of employees who earn more than this average salary:

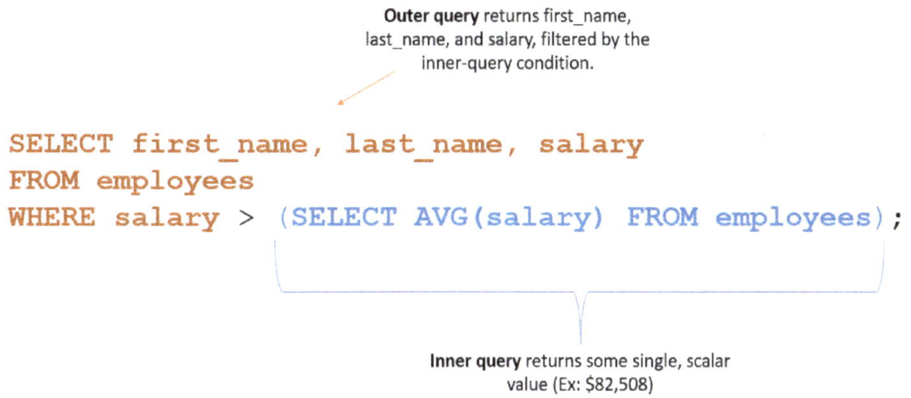

Outer query returns first_name, last_name, and salary, filtered by the inner-query condition.

```
SELECT first_name, last_name, salary
FROM employees
WHERE salary > (SELECT AVG(salary) FROM employees);
```

Inner query returns some single, scalar value (Ex: $82,508)

Figure 5.7: Scalar WHERE subquery explained

> **Note**
>
> Some novice learners may look at this example and wonder, "Why can't I just use `WHERE salary > AVG(salary)`?" Indeed, that would be more straightforward, but unfortunately, SQL does not work this way. This is because aggregate functions such as `AVG`, `MIN`, and `MAX` cannot be used in a `WHERE` clause. Furthermore, we cannot use `HAVING` in this case either, because there is no grouping taking place – hence the need for the subquery.

Non-scalar example

Let's look at a non-scalar example. Suppose that we are using the same dataset as before with the `first_name`, `last_name`, and `salary` fields. We want to return the first name, last name, and salary of employees whose first name begins with the letter `'J'`:

```
SELECT first_name, last_name, salary
FROM employees
WHERE salary > ANY (SELECT salary FROM employees WHERE first_name LIKE
'J%');
```

Let's evaluate this multi-step process:

- **Step 1**: Starting from the innermost query, we select `salary` from the `employees` table where the first name begins with the letter `'J'`. If the row satisfies the subquery condition, it will evaluate `True`, which means that the row is returned in the results. If the row does not satisfy the condition, the row will be filtered out by the outer query's `WHERE` clause. Unlike the scalar value example, this subquery returns multiple rows.

- **Step 2**: Once the interpreter determines which rows will be returned in the inner query, the outer query uses this as the new base dataset. The `WHERE` clause of the outer query filters the subquery to rows where the salary is greater than any row's salary with an employee whose first name starts with `'J'`. To accomplish this, the `ANY` operator identifies any salary from the subquery and filters the entire `employees` table to rows where the salary is greater than those from the subquery.

Subqueries in the HAVING clause

The `HAVING` clause is used to filter the results of a `GROUP BY` query based on conditions involving aggregate functions. The subquery is executed for each group and filters the groups based on the specified condition.

Here are some situations where subqueries in the HAVING clause are useful:

- **Filtering groups based on aggregates**: Subqueries in the HAVING clause are particularly useful when you need to filter groups based on aggregate calculations. For example, you can use a subquery to identify groups where the average order amount exceeds a certain threshold or groups where the count of orders meets specific criteria.

- **Applying conditional filters**: Subqueries in the HAVING clause allow you to apply conditional filters to the grouped results. This is especially handy when you want to include or exclude groups based on certain conditions. For instance, you can use a subquery to filter groups with a maximum value above a specified threshold or groups where a specific condition is met.

- **Comparing aggregates across groups**: Subqueries in the HAVING clause can help compare aggregate values across different groups. You can use a subquery to calculate aggregate values within each group and then compare those values across groups to identify patterns or variances.

Here is an example:

```
SELECT CustomerID, AVG(TotalAmount) AS AverageTotalAmount
FROM Orders
GROUP BY CustomerID
HAVING AVG(TotalAmount) > (SELECT AVG(TotalAmount)
FROM Orders);
```

The subquery (SELECT AVG(TotalAmount) FROM Orders) is used within the HAVING clause to compare the average total amount for each customer with the overall average total amount. It helps filter the results based on the condition specified in the subquery.

Distinguishing common table expressions (CTEs) from subqueries

Many SQL students confuse CTEs with subqueries, so now is a great time to make the distinction between the two. CTEs are also temporary tables typically that are formulated at the beginning of a query and only exist during the execution of the query. This means that CTEs cannot be used in other queries beyond the one in which you are using the CTE.

While CTEs and subqueries are both used in similar circumstances (such as when you need to produce an intermediary result), there are a couple of factors that tip off CTEs:

- They are typically created at the beginning of a query using the WITH operator
- They are followed by a query that queries the CTE

Alternatively, subqueries are a query within a query, nested within one of a query's clauses.

Here is how CTEs are constructed:

```
WITH alias AS ( <Put query here>
)
…. <Query that queries the alias>
```

Here is a more concrete example of using a CTE:

```
WITH customer_totals AS (
SELECT CustomerID, SUM(TotalAmount) AS total_sales
FROM Orders
GROUP BY CustomerID )
SELECT c.CustomerID, c.total_sales, o.avg_order_amount
FROM customer_totals c
JOIN (
SELECT CustomerID, AVG(TotalAmount) AS avg_order_amount
FROM Orders GROUP BY CustomerID ) o
ON c.CustomerID = o.CustomerID;
```

Here's what's happening:

- The CTE is defined using the WITH keyword and given the name customer_totals. Inside the parentheses, the CTE consists of a simple SELECT statement that calculates the total sales for each customer by summing the TotalAmount column of the Orders table. The result is grouped by CustomerID and aliased as total_sales.

- The outer SELECT statement retrieves CustomerID and total_sales from the CTE, as well as the avg_order_amount subquery.

- The FROM clause of the main query references the customer_total CTE directly as the "c" source table.

- The subquery in the JOIN clause calculates the average order amount for each customer The result is grouped by CustomerID and aliased as avg_order_amount. The JOIN condition connects the main query with the subquery using the CustomerID key column.

- The final result set is returned, showing the customer ID, total sales, and average order amount for each customer.

> **Note**
> Like subqueries, CTE tables can also be used in other clauses beyond FROM, such as WHERE and SELECT.

In conclusion, when you approach SQL problems that require filtering, ask yourself if the condition is something that can be hardcoded, or if it requires a calculation. Then, ask yourself if you are filtering to a single scalar value or multiple rows.

Assessment

Consider a table called `Sales` with the `SaleId`, `ProductId`, `SaleDate`, `SaleAmount`, and `CustomerId` columns. Write a SQL query to retrieve `CustomerId` and total `SaleAmount` (aliased as `TotalSaleAmount`) for customers who made at least one purchase with `SaleAmount` over 1,000. The results should be ordered by `TotalSaleAmount` (sum of `SaleAmount`) in descending order.

Answer

Here's the answer:

```
SELECT CustomerId, SUM(SaleAmount) as TotalSaleAmount
FROM Sales
WHERE SaleId IN (
  SELECT SaleId
  FROM Sales
  WHERE SaleAmount > 1000)
GROUP BY CustomerId
ORDER BY TotalSaleAmount DESC;
```

The subquery in the WHERE clause filters `SaleId` values, where `SaleAmount` is greater than 1000. The main query then uses these `SaleId` values to filter the `Sales` table and get `CustomerId` and total `SaleAmount` for these sales. The GROUP BY clause groups the data by `CustomerId`, and the SUM function calculates the total `SaleAmount`. Finally, the ORDER BY clause sorts the result by `TotalSaleAmount` in descending order.

Assessment

Rewrite the previous answer using a CTE instead of a subquery.

Answer

Here's the answer:

```
WITH filtered_sales AS (
SELECT SaleId FROM Sales WHERE SaleAmount > 1000 )
SELECT CustomerId, SUM(SaleAmount) AS TotalSaleAmount
FROM Sales WHERE SaleId IN (
SELECT SaleId FROM filtered_sales)
```

```
GROUP BY CustomerId
ORDER BY TotalSaleAmount DESC;
```

Merging tables with joins

SQL joins are used to combine rows from two or more tables based on a related column between them, providing a complete view of the data. We previously hinted at these related columns as primary keys and foreign keys.

As a refresher, a primary key is a column (or a combination of columns) in a database table that uniquely identifies each row in that table. A foreign key, on the other hand, is a column or a combination of columns in a table that establishes a link or a relationship to the primary key of another table.

As we dive into SQL joins, we will put our knowledge of primary and foreign keys to work!

> **Note**
>
> When discussing SQL joins, we will mostly focus on joining two tables to simplify the concepts. Traditionally, two joined tables are referred to as the left table and the right table.

Inner joins

INNER JOIN selects records that have matching values in both tables. *Figure 5.8* best demonstrates the logic of this join type:

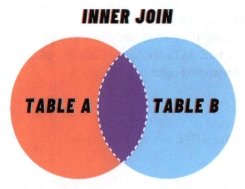

Figure 5.8: Inner join logic

Table A represents the left table and Table B represents the right table. Both tables share a key (primary and foreign key respectively). When performing an inner join, the returned results are the rows that exist in both Table A and Table B.

Let's consider an example where we have two tables, `Orders` and `Customers`:

OrderID	OrderDate	CustomerID
1	2023-06-01	101
2	2023-06-02	102
3	2023-06-03	103
4	2023-06-04	101
5	2023-06-05	102
6	2023-06-06	104
7	2023-06-07	105

CustomerID	CustomerName	City
101	John	London
102	Sarah	Paris
103	Michael	Berlin
106	Emma	Madrid

Figure 5.9: The Orders and Customer tables

We want to list customers from the customer table with their orders from the order table. However, we only want customers who have orders. This is a job for inner join! To begin, we can use the `INNER JOIN` and `ON` keywords to perform an inner join, like so:

```
SELECT Orders.OrderID, Customers.CustomerName
FROM Orders
INNER JOIN Customers ON Orders.CustomerID = Customers.CustomerID;
```

Let's discuss some of this code:

- The first row of code selects the `OrderID` column from the `Orders` table and the `CustomerID` column from the `Customers` table. Since we are dealing with more than one table in this query, we preface every column name with its respective table name, separated by a dot (`.`).
- The second row designates that we are querying from the `Orders` table (see the following *Note* box for more).
- The last row is the meat of the joining process. We call `INNER JOIN` on the `Customers` table (this is because we specified the `Orders` table in the `FROM` clause – yes, you could have done this in reverse). All that's left is describing which fields should be used to perform the inner join.
- We call `ON` on the `Orders` table's `CustomerID` and set it equal to the `Customer` table's `CustomerID` field:

> **Note**
>
> The **ON** keyword is used with all joins. It describes how to connect the two tables by identifying the primary key.

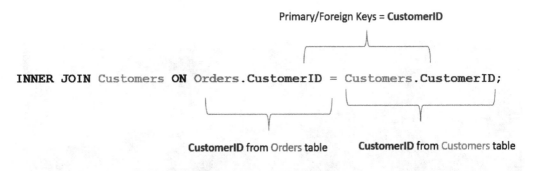

Figure 5.10: Inner join logic

This query returns a list of order IDs, along with the name of the customer who made each order.

> **Note**
>
> As you'll soon learn, joins syntax is often relative. What's considered the left table and right table are completely up to you. Furthermore, the table specified in the FROM clause in the previous example could have been Customers instead of Orders – the results would have been the same since we are using an inner join. In either case, whichever table you don't use in the FROM clause is the table you'll start with in the INNER JOIN clause. Furthermore, you would have to update the following the ON operator. But beware – as you'll see in other join types, the table in the FROM clause will matter.

Novice learners might look at the join syntax and get confused, but there is a pattern that can help you remember it: *A, B, A, B,* or *B, A, B, A.*

Let's see how this applies to our most recent example with the Orders and Customers tables. Notice that the syntax in *Figure 5.10* calls the Customers table in the INNER JOIN clause (which we'll call Table A). Presuming that Orders is Table B, the rest of the code is easy to remember:

```
...ON TableB.Key = TableA.Key;
```

This can also be specified like so:

```
Customers ON Orders.Customer_ID = Customers.Customer_ID;
```

Thus, you should note that either of the following two approaches are correct:

```
OPTION 1:
FROM TableA
INNER JOIN TableB ON TableA.Shared_Key = TableB.Shared_Key;


OPTION 2:
FROM TableB
INNER JOIN TableA ON TableB.Shared_Key = TableA.Shared_Key;
```

Figure 5.11: Same inner join with two different approaches

Left and right join

LEFT JOIN (also known as the LEFT OUTER join) selects all records from the left table and any matching records in the right table:

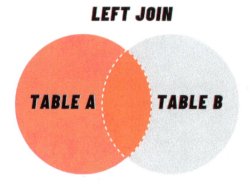

Figure 5.12: Left join logic

If no match is found between the tables, the returned row will show values from Table A, and NULL values from the right table.

RIGHT JOIN (also known as the RIGHT OUTER join) does just the opposite. It returns all of the records from the right table (Table B) and matches the records from the left table (Table A). Any examples where there is no match will result in NULL values from the left table.

If you switched Table A to Table B and Table B to Table A and performed a right join, the results would be the same as if you never switched the tables and performed a left join. How is that for a mind twister?

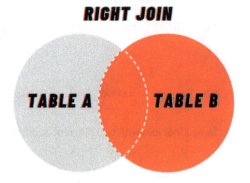

Figure 5.13: Right join logic

Here is the syntax for using `LEFT JOIN` and `RIGHT JOIN`:

```
SELECT column_name(s)
FROM table1
LEFT JOIN table2
ON table1.column_name = table2.column_name;

SELECT column_name(s)
FROM table1
RIGHT JOIN table2
ON table1.column_name = table2.column_name;
```

Again, the `ON` clause defines how to join the two tables.

If we wanted to select all customers and any order information available, we would use `LEFT JOIN`, like this:

```
SELECT Customers.CustomerName, Orders.OrderID
FROM Customers
LEFT JOIN Orders ON Customers.CustomerID = Orders.CustomerID;
```

Suppose we wanted to select all orders and any customer information available; we would use `RIGHT JOIN`:

```
SELECT Orders.OrderID, Customers.CustomerName
FROM Customers
RIGHT JOIN Orders ON Customers.CustomerID = Orders.CustomerID;
```

> **Note**
>
> You might be wondering why right joins exist in the first place since they can be achieved by switching the left and right table designations. The most relevant explanation to consider is query optimization and performance. For example, there are cases where optimizing the query performance can be influenced by the relative sizes of the left and right tables in a join operation.

Full outer join

FULL OUTER JOIN returns all the rows from both tables, regardless of whether there is a match or not. It combines the results of a left outer join and a right outer join. If a row from one table does not have a match in the other table, the result will contain NULL values for the columns of the non-matching table:

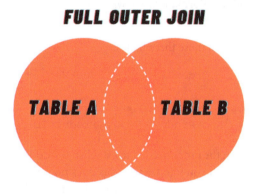

Figure 5.14: Full join logic

Let's look at the syntax:

```
SELECT column_name(s)
FROM table1
FULL OUTER JOIN table2
ON table1.column_name = table2.column_name;
```

As an example, the following query will return the result set with the CustomerName and OrderID columns representing the joined data from the Customers and Orders tables. All rows from both tables are included in the result set, along with the relevant columns – that is, CustomerName and OrderID:

```
SELECT Customers.CustomerName, Orders.OrderID
FROM Customers
FULL OUTER JOIN Orders ON Customers.CustomerID = Orders.CustomerID;
```

SQL joins are an essential feature of the SQL language, and understanding them is a must for any data scientist. Being proficient in joins not only helps you in data manipulation and querying tasks but also proves beneficial in technical interviews since understanding joins is a fundamental part of relational database management.

Multi-table joins

It's common to run into situations where you need to join more than just two tables. Fortunately, the process is the same. You just need to remember what table you're joining to what. Keeping track of this order will ensure you produce the desired results.

Let's consider an example where we have three tables: `Customers`, `Orders`, and `Products`. The `Customers` table contains customer information, the `Orders` table stores order details, and the `Products` table contains product information. We want to retrieve the customer name, order date, and product name for each order.

Here's an example SQL query to join these three tables:

```
SELECT c.CustomerName, o.OrderDate, p.ProductName
FROM Customers c
INNER JOIN Orders o ON c.CustomerID = o.CustomerID
INNER JOIN Products p ON o.ProductID = p.ProductID;
```

The `JOIN` clauses are arranged in the order needed to establish the desired connections between the tables. Note that the order in which you join these tables does not matter as the ultimate goal is to have all three tables joined. This is yet another perk when using inner joins. However, there are cases where the order in which you join the tables would matter. Regard the following example:

```
SELECT c.CustomerName, o.OrderDate, p.ProductName
FROM Customers c
INNER JOIN Orders o ON c.CustomerID = o.CustomerID
LEFT JOIN Products p ON o.ProductID = p.ProductID;
```

In this example, the `Orders` table is joined before the `Products` table, and we use `LEFT JOIN` between the `Orders` and `Products` tables. This ensures that all records from the `Orders` table are included in the result, regardless of whether there is a matching record in the `Products` table. The join condition connects the `Orders` and `Products` tables based on the `ProductID` column.

Assessment

Consider two these two tables:

- `Orders`, with columns for `OrderId`, `CustomerId`, and `OrderDate`
- `Customers`, with columns for `CustomerId`, `FirstName`, `LastName`, and `Country`

Write a SQL query to retrieve all the orders, along with customer details. If a customer doesn't have any orders, include those customers in the result. The result should contain the `OrderId`, `OrderDate`, `CustomerId`, `FirstName`, `LastName`, and `Country` columns.

Answer

Here's the answer:

```
SELECT o.OrderId, o.OrderDate, c.CustomerId, c.FirstName, c.LastName,
c.Country
FROM Customers c
LEFT JOIN Orders o ON c.CustomerId = o.CustomerId;
```

Here, we use `LEFT JOIN` to combine rows from `Customers` and `Orders`. This type of join returns all the rows from the `Customers` table (left table) and the matched rows from the `Orders` table (right table). If no match is found, `NULL` is returned for the columns of the `Orders` table. This ensures that even customers without orders are included in the result.

Calculating window functions

SQL window functions are an additional tool in your toolkit. Unlike aggregate functions, which return a single result per group of rows, **window functions** return a single result for each row, based on the context of that row within a *window* of related rows.

OVER, ORDER BY, PARTITION, and SET

Window functions have the following basic syntax:

```
<function> (<expression>)
OVER (
[PARTITION BY <expression_list>]
[ORDER BY <expression_list>] [ROWS|RANGE <frame specification>])
```

There are a few key concepts to understand here, so let's break them down:

- The `OVER` keyword is what differentiates a window function from a regular function; once you see it, you know you're in window function land. The `OVER` clause defines the window or subset of rows within a query result set that the window function operates on. In short, it provides a way to partition the result set into logical groups and allows the window function to perform calculations or aggregations over those groups. Regard *Figure 5.15* for an illustrative example.

- Inside the `OVER` clause, you can use the `PARTITION BY` keyword to break the data into separate windows. It divides the result set into distinct partitions or groups based on one or more columns. The window function is then applied separately to each partition, allowing calculations or aggregations to be performed within each distinct group.

> **Note:**
>
> PARTITION BY is an optional operator and is not needed to perform a window function. Only OVER is needed to initialize a window function. However, it is PARTITION BY that allows you to perform some operation over grouped categories, so it's common to see it used with OVER. Without PARTITION BY, a query will consider the entire result set as a single partition. Lastly, ORDER BY is optional, but it is used to sort the data within those windows.

That seems like a lot, so let's go over some examples. Suppose we have the following dataset:

```
Month | Year | State | Revenue ------------------------------------
January | 2022 | New York | 45000
February| 2022 | New York | 47000
March | 2022 | New York | 49000
January | 2022 | Texas | 52000
February| 2022 | Texas | 54000
March | 2022 | Texas | 55000
January | 2023 | New York | 50000
February| 2023 | New York | 52000
March | 2023 | New York | 54000
January | 2023 | Texas | 60000
February| 2023 | Texas | 61000
March | 2023 | Texas | 62000
```

Say that we wanted to group our results by state, and then within those state groupings, show the average revenue, ordered by year. OVER and PARTITION BY allows us to create windows of rows based on a field (in this case, State). We created these windows so that we could calculate the average revenue for those windows. ORDER BY simply allows us to organize the results in those windows. Let's take a look at an example.

Here is the query:

```
SELECT Year, State, Revenue
AVG(Revenue) OVER (
PARTITION BY State ORDER BY Year)
AS AverageRevenue
FROM SalesTable;
```

Let's assume there are three states instead of two. Here are the hypothetical results:

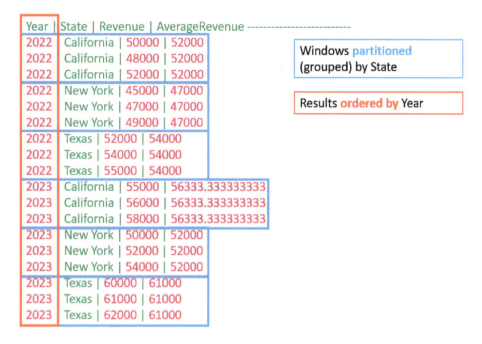

```
Year | State | Revenue | AverageRevenue ---------------------------
2022  California | 50000 | 52000
2022  California | 48000 | 52000
2022  California | 52000 | 52000
2022  New York | 45000 | 47000
2022  New York | 47000 | 47000
2022  New York | 49000 | 47000
2022  Texas | 52000 | 54000
2022  Texas | 54000 | 54000
2022  Texas | 55000 | 54000
2023  California | 55000 | 56333.333333333
2023  California | 56000 | 56333.333333333
2023  California | 58000 | 56333.333333333
2023  New York | 50000 | 52000
2023  New York | 52000 | 52000
2023  New York | 54000 | 52000
2023  Texas | 60000 | 61000
2023  Texas | 61000 | 61000
2023  Texas | 62000 | 61000
```

Windows **partitioned** (grouped) by State

Results **ordered by** Year

Figure 5.15: Window functions demonstrated

Let's review how this table is derived:

- The SELECT clause specifies the columns to be included in the result set: Year, State, Revenue, and the calculated AverageRevenue.

- The FROM clause specifies the SalesTable column from which the result set will be derived.

- The AVG function is used as a window function with the OVER clause to calculate the average revenue within each partition.

- The PARTITION BY clause is used to partition the data by the State column. This means that the data will be grouped and processed separately for each distinct state.

- The ORDER BY clause is used to order the rows within each partition by the Year column. This determines the sequence in which the window function is applied within each state group.

It is the combination of OVER, PARTITION BY, and ORDER BY that makes window functions so powerful. One common use of a window function is to calculate running totals. The SUM() function can be used as a window function to achieve this.

Suppose we have a table called `employees` with the `employee_id`, `first_name`, `last_name`, `salary`, and `department_id` columns, and we want to calculate a running total of salaries within each department. We could use the following query:

```
SELECT employee_id, first_name, last_name, salary, department_id,
       SUM(salary) OVER (PARTITION BY department_id ORDER BY employee_
id) as running_total
FROM employees;
```

This query returns the `running_total` value of `salary` for each row, totaled (summed) over all rows with the same `department_id` and an `employee_id` value less than or equal to the current row.

Window functions shine when paired with SQL functions that are often calculated in windows. This includes LAG, LEAD, ROW_NUMBER, RANK, DENSE_RANK, and NTILE.

LAG and LEAD

LAG is an analytic function in SQL that provides access to a previous row within a result set. It allows us to retrieve the value of a column from the preceding row, enabling us to compare and compute values based on the previous row's data. The syntax is as follows:

```
LAG(column, offset, default) OVER (PARTITION BY partition_clause ORDER
BY order_clause)
```

Let's review what all this stuff means:

- `column` is the column from which we want to retrieve the previous row's value.
- `offset` specifies the number of rows to look back. It is an optional parameter, with a default value of 1.
- `default` is an optional parameter that sets a default value to return if no previous row is found.
- `PARTITION BY` divides the result set into partitions or groups based on specified columns.
- `ORDER BY` determines the order of rows within each partition.

Let's consider a table called "`sales`" with the `order_id`, `order_date`, and `revenue` columns. Say we want to retrieve the previous order's revenue for each order; we would do something like this:

```
SELECT order_id, order_date, revenue,
LAG(revenue) OVER (ORDER BY order_date) AS previous_revenue
FROM sales;
```

This query retrieves `order_id`, `order_date`, and `revenue`, and the previous order's revenue using LAG(). The result set will include the columns from the sales table, along with an additional column named `previous_revenue` containing the revenue from the preceding order.

LEAD is another analytic function in SQL that provides access to a subsequent row within a result set. It allows us to retrieve the value of a column from the following row, enabling us to perform calculations and comparisons based on the subsequent row's data. LEAD and LAG share the same syntax, with the only difference being the function name itself (LEAD or LAG).

Let's continue with our "sales" table example. Suppose we want to calculate the difference in revenue between consecutive orders:

```
SELECT order_id, order_date, revenue,
LEAD(revenue) OVER (ORDER BY order_date) - revenue AS revenue_
difference
FROM sales;
```

This query retrieves order_id, order_date, and revenue, and calculates the revenue difference between consecutive orders. The result set will include the columns from the sales table along with an additional column named revenue_difference representing the difference in revenue.

Assessment

Consider a table called employees with the employee_id, employee_name, and hire_date columns. Retrieve the previous hire date for each employee, ordered by hire_date.

Answer

Here's the answer:

```
SELECT employee_id, employee_name, hire_date,
LAG(hire_date) OVER (ORDER BY hire_date) AS previous_hire_date FROM
employees
```

The LAG function with the ORDER BY clause retrieves the value from the previous row within the result set based on the specified column (hire_date). By ordering the rows by hire_date, we ensure that LAG looks back at the hire date of the preceding employee for each row. The result is a dataset that includes the employee details, along with the hire date of the employee's predecessor.

Assessment

Consider a table called orders with the order_id, order_date, and revenue columns. Calculate the revenue difference between consecutive orders, ordered by order_date.

Answer

Here's the answer:

```
SELECT order_id, order_date, revenue,
LEAD(revenue) OVER (ORDER BY order_date) - revenue AS revenue_
difference
FROM orders;
```

The LEAD function with the ORDER BY clause retrieves the value from the next row within the result set based on the specified column (order_date). By ordering the rows by order_date, we ensure that the LEAD function looks ahead at the revenue of the subsequent order for each row. Subtracting the current order's revenue from the subsequent order's revenue gives us the revenue difference. The result is a dataset that includes the order details, along with the revenue difference between consecutive orders.

ROW_NUMBER

ROW_NUMBER is an analytic function in SQL that assigns a unique number to each row within a result set. It generates a sequential integer starting from 1 for the first row and increments by one for each subsequent row. This function is very useful for exercises such as ranking, detecting duplicates, or pagination. Here is the syntax:

```
ROW_NUMBER() OVER (
PARTITION BY partition_clause ORDER BY order_clause)
```

Let's consider a table called students with the student_id, student_name, and exam_score columns. We want to assign a unique row number to each student based on their exam score, ordering them in descending order. We would do this like so:

```
SELECT student_id, student_name, exam_score,
ROW_NUMBER() OVER (ORDER BY exam_score DESC) AS row_number FROM
students;
```

This query retrieves student_id, student_name, and exam_score, and assigns a unique row number to each student using ROW_NUMBER. The result set will include the columns from the students table, along with an additional column named row_number containing the sequential numbers.

RANK and DENSE_RANK

RANK is an analytic function in SQL that assigns a unique rank to each row within a result set based on the specified criteria. It allows us to determine the ranking position of a row in comparison to others, considering ties and skipping ranks if necessary.

Similarly, **DENSE_RANK** is another analytic function in SQL that assigns a unique rank to each row within a result set based on the specified criteria. Unlike RANK, it does not skip ranks when there are ties. Instead, it assigns consecutive ranks to tied rows.

Here is the syntax for both:

```
[RANK() or DENSE_RANK()] OVER (PARTITION BY partition_clause ORDER BY
order_clause)
```

Let's consider a table called `students` with the `student_id`, `student_name`, and `exam_score` columns. We want to rank the students based on their exam scores, ordering them in descending order with the highest scorers on top. Here's an example query using RANK:

```
SELECT student_id, student_name, exam_score,
RANK() OVER (
ORDER BY exam_score DESC) AS rank FROM students;
```

This query retrieves `student_id`, `student_name`, and `exam_score`, and assigns a unique rank to each student based on their exam score. The scores are presented in descending order.

However, if we replace RANK with DENSE_RANK, the results will be different if there are ties. RANK leaves gaps in the ranking sequence when there are ties, while DENSE_RANK assigns consecutive ranks to tied rows without any gaps.

For example, say that two of the students received a score of 98, and this is the top score in the data. Using DENSE_RANK, they will both be assigned a rank of 1, and the next highest scorer(s) will receive a rank of 2. Using RANK, the two students with the score of 98 will still receive a rank of 1 but the second-highest scorer will be given a rank of 3. This is because RANK skips ranks if there are ties.

Here is an example of using DENSE_RANK:

student_id	student_name	exam_score	dense_rank
2	Emily	92	1
4	Michael	92	1
3	Sarah	88	2
1	John	85	3
5	Sophia	85	3

Figure 5.16: DENSE_RANK output

Here is an example of using RANK:

Figure 5.17: RANK output

Notice that Michael's rank is higher (2 instead of 3) while using DENSE_RANK. In short, if you're taking a class that's making you feel a little *dense*, you'd probably prefer your teacher to rank you using the DENSE_RANK method!

Assessment

Given a table called Sales with the SaleId, ProductId, SaleDate, SaleAmount, and EmployeeId columns, write a SQL query to find the total sales for each EmployeeId, along with the rank value of each employee in terms of total sales. The rank should be in descending order of total sales, with the employee having the highest total sales ranked first.

Answer

Here's the answer:

```
SELECT EmployeeId, SUM(SaleAmount) OVER (PARTITION BY EmployeeId) AS
TotalSales,
       RANK() OVER (ORDER BY SUM(SaleAmount) OVER (PARTITION BY
EmployeeId) DESC) AS SalesRank
FROM Sales;
```

This query introduces two window functions:

- SUM(SaleAmount) OVER (PARTITION BY EmployeeId) calculates the total sales for each employee

- RANK() OVER (ORDER BY SUM(SaleAmount) OVER (PARTITION BY EmployeeId) DESC) assigns a rank to each employee, based on their total sales, in descending order.

Using date functions

Date functions in SQL are used for manipulating date data types, and they are essential for performing operations such as calculating differences between dates, extracting date parts, and formatting dates. While specific functions may vary slightly between different SQL databases, most databases support a core set of date functions.

Let's review several of the most common functions:

- **NOW**: The NOW function returns the current date and time:

```
SELECT NOW() AS 'Current Date and Time';
```

- **CURDATE**: The CURDATE function returns the current date:

```
SELECT CURDATE() AS 'Current Date';
```

- **DATE_ADD**: The DATE_ADD function is used to add or subtract date parts. The parameters of this function include a date value, an INTERVAL value, and an interval size. For example, if you are looking to add 2 days to each row for the date column, you would write the DAY date value. This is followed by INTERVAL and an interval value, which is 2 in this case. However, if you use a negative value for the interval, it will subtract from the date. If you want to calculate the date 30 days from now, use DATE_ADD:

```
SELECT DATE_ADD(date_column, INTERVAL 2 DAY) AS '2 Days Later';
```

DATEDIFF: DATEDIFF calculates the difference between two dates. Suppose we have a table called orders with the order_id, product_id, and order_date columns and we want to calculate how long it's been since each order was placed. We would do this:

```
SELECT DATEDIFF(NOW(), order_date) AS 'Days Since Order'
FROM orders;
```

SQL date functions are a crucial part of performing complex date manipulations and calculations.

Approaching complex queries

Writing complex SQL queries can be a challenging task, especially when dealing with multiple tables, complex filtering conditions, and intricate calculations. However, by following a step-by-step approach, you can break down the problem into smaller, manageable parts and gradually build up to the final query.

Here are some systematic guidelines on how to approach complex queries:

- **Step 1 – define the objective**: Begin by clearly defining the objective of your query. What specific information are you trying to retrieve or calculate? What is the desired output?

- **Step 2 – identify the tables**: Determine which tables contain the necessary data for your objective and identify their respective keys. This helps us identify our starting point. If multiple tables are involved, consider the relationships between them and how they should be joined. Determine the key(s) in each table.

- **Step 3 – determine the filtering criteria**: Identify the filtering criteria needed to narrow down the dataset. Determine which conditions should be applied to limit the rows returned. Consider both the explicit conditions (such as WHERE clauses) and any implicit conditions that may be required. Which table is being filtered? Is it the inner or outer query?

- **Step 4 – start with simple joins**: If your query involves multiple tables, start by performing simple joins between the relevant tables. Determine which table will be on the left, and how it will be joined with other tables. Begin with the primary relationship and gradually add additional join conditions as needed.

- **Step 5 – incorporate aggregates**: If your query requires aggregating data, determine the appropriate aggregate function(s). Consider if any grouping or partitioning is necessary to aggregate data at the desired level. For each aggregate, be sure to consider what level your aggregations should take place. Is it over the entire dataset? Is it by grouping? Is it over specific window segments? If grouped, consider the single-valued grouping rule.

- **Step 6 – evaluate subqueries and CTEs**: If the complexity of your query demands it, consider incorporating subqueries or CTEs to handle calculations, temporary views, or filter results. Review the granularity needs of each aggregate function for opportunities to use subqueries.

- **Step 7 – review**: Return to *Step 1* to confirm that you have achieved the objective.

By following this step-by-step approach, you can tackle complex SQL queries more effectively. Although there is room to adjust the order of these steps, it is recommended to stick as closely to this framework as possible, and do not be afraid to walk through these steps out loud!

Assessment

You are working with a database that contains three tables. The interviewer has asked you to retrieve the total order amount for each customer, along with the product details of their most expensive order. Your output should be CustomerID, CustomerName, MaxOrderAmount, and TotalOrderAmount. Here are the table's contents:

- Customers: CustomerID, CustomerName, CustomerAddress, and CustomerEmail

- Orders: OrderID, CustomerID, OrderDate, and OrderAmount

- Products: ProductID, ProductName, ProductPrice, and ProductCategory

Process and answer

Here's the process:

1. **Define the objective**: Our objective is to retrieve the total order amount for each customer and include the product details of their most expensive order. Based on the instruction, we need to return `CustomerID`, `CustomerName`, and `ProductName`, as well as the calculated fields, `TotalOrderAmount` and `MaxOrderAmount`. Although we don't know where all of this information is coming from at this point, you can include it in the query as we know this is where we want to be at the end of the query development. Be sure to name any calculated fields exactly as instructed.

   Here is the query thus far:

   ```
   SELECT CustomerID, CustomerName, SUM(OrderAmount) AS
   TotalOrderAmount, ProductName, ... AS MaxOrderAmount ...
   ```

   We will calculate `MaxOrderAmount` later. While `TotalOrderAmount` is an aggregate we need for all unique customers, `MaxOrderAmount` is needed for only the most expensive orders.

2. **Identify the tables**: Order information will come from the `Orders` table. Customer information will come from the `Customers` table. Product details will come from the `Products` table.

3. **Determine the filtering criteria**: We don't have any specific filtering criteria for this objective. We want to retrieve information for all customers.

4. **Start with simple joins**: Based on our objective, we need total order information from the `Orders` table. Since this is for each customer, we will want all customers from the `Customers` table. This constitutes an inner join with the `Customers` table and the `Orders` table. Their shared key is the `CustomerID` field. Since we are dealing with more than one table, we will need to provide table aliases before every field name, separated by a dot.

   Now, let's review the query again:

   ```
   SELECT c.CustomerID, c.CustomerName, SUM(o.OrderAmount) AS
   TotalOrderAmount, ProductName, ... AS MaxOrderAmount
   FROM Customers c
   INNER JOIN Orders o ON c.CustomerID = o.CustomerID;
   ```

   Since we will also need product details, we will join the `Products` table to our already joined `Customers` and `Orders` table. Using an inner join yet again confirms that all products with a customer and order will be returned. The `Customers` table does not share an ID with `Products`, but it does with `Orders`, so we will use it: `ProductID`.

   Let's see the query with this new information:

   ```
   SELECT c.CustomerID, c.CustomerName, SUM(o.OrderAmount) AS
   TotalOrderAmount, p.ProductName, ... AS MaxOrderAmount
   FROM Customers c
   ```

```
INNER JOIN Orders o ON c.CustomerID = o.CustomerID
INNER JOIN Products p ON o.ProductID = p.ProductID;
```

5. **Incorporate aggregates**: We've already used SUM on OrderAmount to derive TotalOrderAmount. Since this value should aggregate for each customer, we need to use GROUP BY on CustomerID and CustomerName. We will also need TotalOrderAmount for each product since we want product details for the product with MaxOrderAmount.

   Let's review the updated query at this point:

```
SELECT c.CustomerID, c.CustomerName, SUM(o.OrderAmount) AS
TotalOrderAmount, p.ProductName, ... AS MaxOrderAmount
FROM Customers c
INNER JOIN Orders o ON c.CustomerID = o.CustomerID
INNER JOIN Products p ON o.ProductID = p.ProductID;
GROUP BY c.CustomerID, c.CustomerName, p.ProductName;
```

   We can now include the product details of each customer's most expensive order, so we'll incorporate a subquery.

6. **Evaluate subqueries and CTEs**: Lastly, we need to calculate MaxOrderAmount. If you think the answer is just calling MAX(o.OrderAmount) AS MaxOrderAmount in the SELECT clause, think again! We must be mindful of granularity. Using MAX on OrderAmount would provide the maximum OrderAmount value for each unique combination of CustomerID, CustomerName, and ProductName, but this isn't our objective – our objective is to return the maximum order amount that is equal to the maximum order amount among all orders. As this is a filtering task, we will use the WHERE clause. This sounds like a filtering exercise. (Note: this step could have been achieved in *Step 3*, but for demonstration purposes, we will do it here.) We include the WHERE CustomerID = c.CustomerID condition in the subquery to ensure that the subquery correlates with the outer query by matching the CustomerID values.

   We are now ready to implement the subquery, like so:

```
SELECT c.CustomerID, c.CustomerName, SUM(o.OrderAmount) AS
TotalOrderAmount, p.ProductName, o.OrderAmount AS MaxOrderAmount
FROM Customers c
INNER JOIN Orders o ON c.CustomerID = o.CustomerID
INNER JOIN Products p ON o.ProductID = p.ProductID
WHERE o.OrderAmount = (
SELECT MAX(OrderAmount)
FROM Orders
WHERE CustomerID = c.CustomerID)
GROUP BY c.CustomerID, c.CustomerName, p.ProductName;
```

7. **Review**: Review your query to ensure that you've achieved all of the necessary objectives!

Note that we could have achieved the same results using a CTE instead of the subquery to calculate the max order amount for each customer. We would just need to join it onto our table and filter to where the order amount is equal to the max amount. This is what we would do instead:

```
WITH MaxOrderAmounts AS (
SELECT CustomerID, MAX(OrderAmount) AS MaxOrderAmount
FROM Orders
GROUP BY CustomerID )
SELECT c.CustomerID, c.CustomerName, SUM(o.OrderAmount) AS
TotalOrderAmount, p.ProductName, o.OrderAmount AS MaxOrderAmount
FROM Customers c
INNER JOIN Orders o ON c.CustomerID = o.CustomerID
INNER JOIN Products p ON o.ProductID = p.ProductID
INNER JOIN MaxOrderAmounts moa ON c.CustomerID = moa.CustomerID
WHERE o.OrderAmount = moa.MaxOrderAmount
GROUP BY c.CustomerID, c.CustomerName, p.ProductName, o.OrderAmount;
```

As you can see, breaking down the problem into smaller steps and gradually building up the query will help you approach complex scenarios with greater confidence and produce accurate results. Remember to practice and experiment with different techniques to further enhance your SQL query writing skills.

Summary

In this chapter, we learned the basics of databases and SQL, which are topics that many data scientists encounter in interviews. In fact, as a data scientist, you will almost certainly be quizzed on this topic during interviews. We touched on basic querying concepts, subqueries, joins, window functions, evaluation order, aggregation, filtration, and how to approach complex problems. However, SQL is yet another topic that commands an entire book on its own.

Rest assured that in most cases, there is more than one way to solve a problem, but there are often limited optimal ways to do so. Thus, be sure to spend adequate time practicing the concepts discussed in this chapter. Try not to memorize queries; instead, familiarize yourself with the common use cases explained in this chapter. Follow the aforementioned steps to break down complex problems, but be aware that the order of these steps is not set in stone. Once mastered, you will be able to identify the right query for any occasion!

In the next chapter, we will look at shell and bash scripting with Linux.

6

Scripting with Shell and Bash Commands in Linux

In this brief chapter, we'll delve into shell and Bash scripting with Linux, covering basic navigation control statements, functions, data processing and pipelines, and database operations. Additionally, you'll learn how to leverage the `cron` command for task scheduling and, importantly, how to run Python programs from the command line.

Although the likelihood of being tested on Linux commands during a data science interview is rare, you'll be better prepared to utilize data science-adjacent technologies that leverage the command line. In this chapter, we will cover the following topics:

- Introduction to operating systems
- Navigating system directories
- Filing and directory manipulation
- Scripting with Bash
- Introducing control statements
- Creating functions
- Processing data and pipelines
- Using cron

Introducing operating systems

An **operating system** (**OS**) is a software program that acts as an intermediary between computer hardware and user applications. You're probably familiar with Windows, Android, and iOS, which are all different types of operating systems with their own unique features and applications.

Linux is an open source OS known for its Unix-like architecture, allowing users to configure and modify the system according to their specific needs. Like other Unix-based systems, it arranges files and directories in a hierarchical structure. The root directory is at the very top of this hierarchy, denoted by a forward slash (/).

The **root directory** is the top-level directory in an OS filesystem's tree-like hierarchy and is the starting point for all other directories and files. For example, if you see a file path such as /home/user/file.txt, the leading forward slash indicates that it is referencing a location relative to the root directory. That location is a file called file.txt in the user directory, under the home directory, under the root directory.

In addition to /home, there are other directories within the root directory, such as /home, /usr, and /etc, each serving a specific purpose. Learning how to navigate these files in the command line will put you ahead of the curve when it comes to expediting workflows and navigating other technologies.

The **command-line interface** (**CLI**) or **shell** is a text-based interface in Linux that allows users to interact with the computer by entering commands. In the rest of this chapter, we will learn how to navigate the Linux OS and its directories using Bash scripting and shell commands in the CLI!

Navigating system directories

One of the foundational aspects of working in a Linux environment is the ability to navigate the file structure and directories from the command line.

If you're familiar with a filesystem on any computer, you're already familiar with this concept. For example, a Windows OS might have a directory (folder) called Desktop, Pictures, Downloads, or Documents. These are all directories. *Figure 6.1* shows an example directory called Physics, which has three text files and a directory called Assignments.

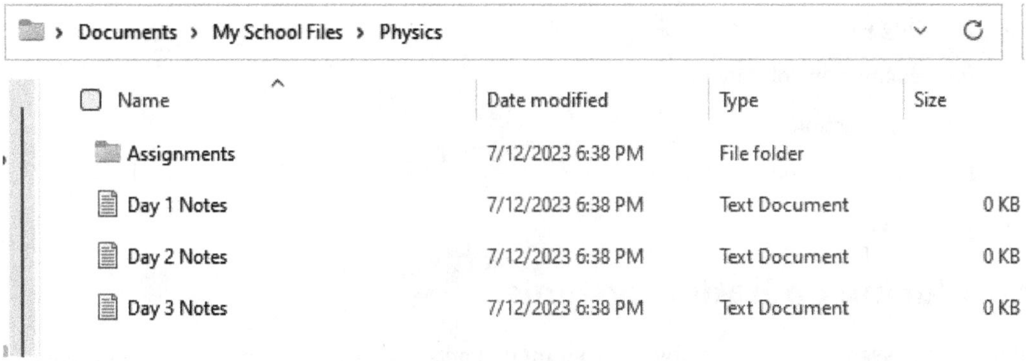

> Documents > My School Files > Physics			
Name	Date modified	Type	Size
Assignments	7/12/2023 6:38 PM	File folder	
Day 1 Notes	7/12/2023 6:38 PM	Text Document	0 KB
Day 2 Notes	7/12/2023 6:38 PM	Text Document	0 KB
Day 3 Notes	7/12/2023 6:38 PM	Text Document	0 KB

Figure 6.1: Example directory titled Physics

A **directory** is a folder, virtual box, container, or organizational structure used to hold and organize files and other directories. *Figure 6.1* illustrates the UI program that allows everyday Windows users to navigate their filesystems. However, the CLI, shown in *Figure 6.2*, enables us to navigate and automate file management using commands.

Figure 6.2: Windows CLI example

Introducing basic command-line prompts

To begin learning how to use the CLI, let's take a look at some basic examples. These are commands that you will use throughout your file exploration journey:

- pwd (print working directory): This command prints the full pathname of the current working directory to the terminal (for example, /home/user/). If you ever get lost in the terminal, pwd is your compass.

- ls (lists): This command lists all files and directories in the current working directory.

- cd <directory_name>: This changes the current working directory to directory_name if it exists in the current directory.

- cd ..: This command navigates up one directory level (note that .. is a separate command on its own, so be sure to distinguish it by leaving a space between cd and ..).

- cd: Without any argument, this command will take you back to your home directory.

- cd -: This command will take you to the previous directory you were in.

Figure 6.3 demonstrates how to use these commands using JSLinux, a Linux OS emulator:

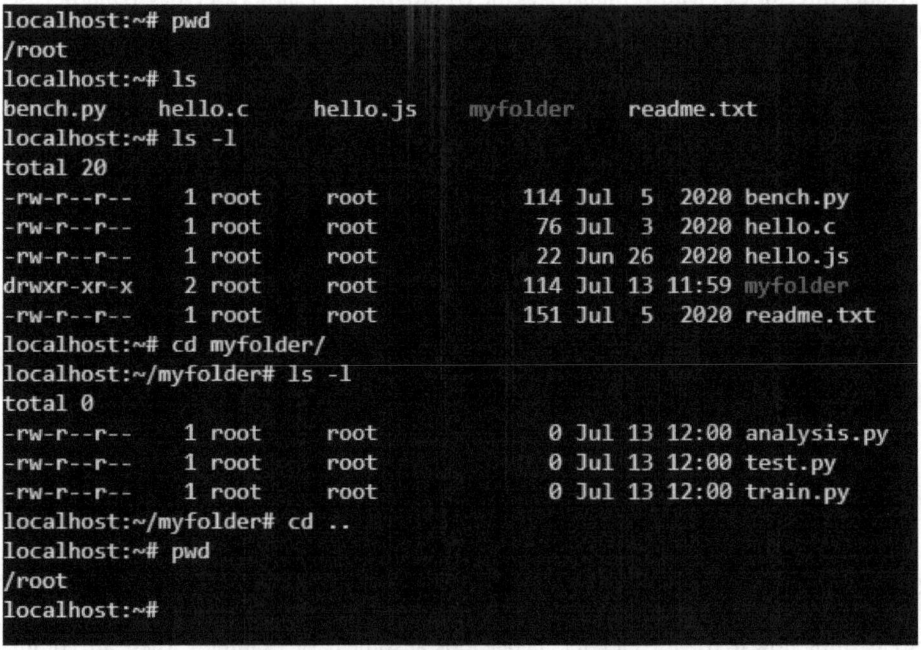

```
localhost:~# pwd
/root
localhost:~# ls
bench.py     hello.c      hello.js     myfolder      readme.txt
localhost:~# ls -l
total 20
-rw-r--r--     1 root      root            114 Jul  5  2020 bench.py
-rw-r--r--     1 root      root             76 Jul  3  2020 hello.c
-rw-r--r--     1 root      root             22 Jun 26  2020 hello.js
drwxr-xr-x     2 root      root            114 Jul 13 11:59 myfolder
-rw-r--r--     1 root      root            151 Jul  5  2020 readme.txt
localhost:~# cd myfolder/
localhost:~/myfolder# ls -l
total 0
-rw-r--r--     1 root      root              0 Jul 13 12:00 analysis.py
-rw-r--r--     1 root      root              0 Jul 13 12:00 test.py
-rw-r--r--     1 root      root              0 Jul 13 12:00 train.py
localhost:~/myfolder# cd ..
localhost:~# pwd
/root
localhost:~#
```

Figure 6.3: Basic Linux commands in action

Note that the `ls -l` command may use the `-l` flag. In Linux, **flags** are *command modifiers*, used to modify the behavior of command-line utilities. They provide additional instructions or settings to a command, allowing users to customize how a command operates, and are typically represented by a hyphen (`-`) followed by a single character or a word. Here, the `-l` flag modifies the output of `ls` to print more comprehensive details and the format of the directory contents.

Understanding directory types

In Linux, there are two methods for accessing directory paths:

- **Absolute paths**: Absolute paths specify the location of a file or a directory from the root directory. They always start with a forward slash – for example, `/home/user/data/file.txt`.

- **Relative paths**: Relative paths specify the location of a file or a directory relative to the current directory. For example, if your current directory is `/home/user/data/`, and you want to navigate to the `/home/user/data/project1/` directory, you could use the following:

```
cd project1
```

The `project1` directory will be interpreted as a relative path to the current working path.

Understanding how to use both path types can help you navigate the filesystem more efficiently. Also, to expedite navigation and using paths, you can leverage the auto-completion feature by hitting the *Tab* key after typing a few characters.

Now that you're familiar with basic navigation, here are some advanced techniques:

- `pushd directory_name` and `popd`: These commands allow you to work with a stack of directories. `pushd` adds a directory to the stack and navigates to it. `popd` removes the top directory from the stack and navigates to it. This can be very useful when you're working with multiple directories and need to switch between them frequently.

- `find`: This is a powerful command to search for files or directories based on criteria such as name, size, and modification time. For instance, `find /home/user -name "file.txt"` will search for a file named `"file.txt"` in the `/home/user` directory and its subdirectories.

You often deal with numerous files and complex directory structures as a data scientist. Command-line navigation, thus, is a vital skill to master. It serves as a stepping stone to more advanced topics such as file and directory manipulation, Bash scripting, cron jobs, and using Python from the command line.

Assessment

Consider that you are currently in the `/home/user/project/dataset1/` directory and you want to change to the `/home/user/project/dataset2/` directory. Using only a single command that includes a relative path, how would you achieve this?

Answer

```
cd ../dataset2/
```

This command navigates up one level to the `project` directory using `../`, and then into the `dataset2` directory. The `../` part is a special directory name, meaning the parent of the current directory, so it always refers to the directory above. The concept of relative path is used here, where the path provided is relative to the current directory.

Assessment

A data scientist is working on a Linux machine. They are in the middle of a complicated data processing task and have navigated to multiple different directories. Now, they want to confirm their current directory within the filesystem. Which command should they use?

Answer

```
pwd
```

The pwd (print working directory) command is used to display the full pathname of the current directory. It is a built-in command in Unix/Linux shells that prints the full pathname to the terminal, which helps users to confirm their current location within the filesystem's hierarchy.

Filing and directory manipulation

Managing files and directories is a fundamental skill when working in a Unix-based environment. As a data scientist, you'll frequently need to create, delete, move, and copy files and directories. Knowing how to use these commands in your daily activities may become a core skill, depending on the systems you are using. However, in a technical interview, these topics might occasionally come up. Therefore, we will only quickly review a few core operations here.

The following list will explain these operations and discuss how to manipulate file and directory contents:

- **Creating files**: To create a new file, use the touch command followed by the name of the file you want to create. For instance, to create a file named analysis.py, you would use the following command:

```
touch analysis.py
```

- **Creating directories**: To create a new directory, use the mkdir command. For example, to create a directory named new_data, use the following:

```
mkdir new_data
```

- **Removing files**: To remove a file, use the rm command. Remember to use this command carefully though, as deleted files cannot be recovered. So, the following example will permanently delete the analysis.py file:

```
rm analysis.py
```

- **Removing directory**: To remove a directory, use rmdir, like so:

```
rmdir new_data
```

Keep in mind that rmdir can only remove an empty directory. To delete a directory and its contents, use the rm command with the -r (recursive) flag:

```
rm -r old_data
```

Here is an example of these commands in action:

```
localhost:~/myfolder# pwd
/root/myfolder
localhost:~/myfolder# ls -l
total 0
-rw-r--r--    1 root      root         0 Jul 13 12:00 analysis.py
-rw-r--r--    1 root      root         0 Jul 13 12:00 test.py
-rw-r--r--    1 root      root         0 Jul 13 12:00 train.py
localhost:~/myfolder# touch experiment1.py
localhost:~/myfolder# ls -l
total 0
-rw-r--r--    1 root      root         0 Jul 13 12:00 analysis.py
-rw-r--r--    1 root      root         0 Jul 13 12:12 experiment1.py
-rw-r--r--    1 root      root         0 Jul 13 12:00 test.py
-rw-r--r--    1 root      root         0 Jul 13 12:00 train.py
localhost:~/myfolder# mkdir predictions
localhost:~/myfolder# ls -l
total 4
-rw-r--r--    1 root      root         0 Jul 13 12:00 analysis.py
-rw-r--r--    1 root      root         0 Jul 13 12:12 experiment1.py
drwxr-xr-x    2 root      root        37 Jul 13 12:13 predictions
-rw-r--r--    1 root      root         0 Jul 13 12:00 test.py
-rw-r--r--    1 root      root         0 Jul 13 12:00 train.py
localhost:~/myfolder# rm analysis.py
```

```
localhost:~/myfolder# ls -l
total 4
-rw-r--r--    1 root      root         0 Jul 13 12:12 experiment1.py
drwxr-xr-x    2 root      root        37 Jul 13 12:13 predictions
-rw-r--r--    1 root      root         0 Jul 13 12:00 test.py
-rw-r--r--    1 root      root         0 Jul 13 12:00 train.py
localhost:~/myfolder# rmdir predictions/
localhost:~/myfolder# ls -l
total 0
-rw-r--r--    1 root      root         0 Jul 13 12:12 experiment1.py
-rw-r--r--    1 root      root         0 Jul 13 12:00 test.py
-rw-r--r--    1 root      root         0 Jul 13 12:00 train.py
localhost:~/myfolder#
```

Figure 6.4: Creating and removing files and directories

- **Moving and renaming files and directories**: The mv command serves two purposes:

  - Moving and renaming files

  - Moving and renaming directories

  Here is the syntax:

  ```
  mv /path/to/source /path/to/destination
  ```

  For example, to rename a file, use this:

  ```
  mv oldname.txt newname.txt
  ```

Here is an example of us moving a file called `experiment1.py` from the `myfolder` directory to the `experiments` directory, located one directory above.

```
localhost:~# ls -l
total 24
-rw-r--r--    1 root      root            114 Jul  5  2020 bench.py
drwxr-xr-x    2 root      root             37 Jul 13 12:21 experiments
-rw-r--r--    1 root      root             76 Jul  3  2020 hello.c
-rw-r--r--    1 root      root             22 Jun 26  2020 hello.js
drwxr-xr-x    2 root      root            117 Jul 13 11:59 myfolder
-rw-r--r--    1 root      root            151 Jul  5  2020 readme.txt
localhost:~# cd myfolder/
localhost:~/myfolder# ls -l
total 0
-rw-r--r--    1 root      root              0 Jul 13 12:12 experiment1.py
-rw-r--r--    1 root      root              0 Jul 13 12:00 test.py
-rw-r--r--    1 root      root              0 Jul 13 12:00 train.py
localhost:~/myfolder# mv experiment1.py ../experiments/
localhost:~/myfolder# cd ../experiments/
localhost:~/experiments# ls -l
total 0
-rw-r--r--    1 root      root              0 Jul 13 12:12 experiment1.py
localhost:~/experiments#
```

Figure 6.5: Moving files between directories

- **Searching within a file for a specific pattern:** The `grep` command searches the file for a specific pattern and prints the matching lines. It's an invaluable tool for searching through large amounts of data. For example, the following command will print out every line in data. csv that includes the string `'San Francisco'`:

```
grep 'San Francisco' data.csv
```

This function also accepts the recursive flag (`-r`), which will allow you to search recursively through a directory. If we modify our previous example, we are now able to search for San Francisco in each file in the `data` directory:

```
grep -r 'San Francisco' /home/user/data
```

As a data scientist, you may use the command line to navigate and manage Python scripts for your machine learning projects or to create data pipelines. These skills will undoubtedly come in handy.

Assessment

You're in a directory that contains a large number of files. You're interested in finding all files that contain the word ERROR in their content. Which command would you use?

Answer

```
grep -r 'ERROR' .
```

The `grep` command is used to search for a specific pattern in file contents. The `-r` option tells `grep` to read all files under each directory, recursively. The `.` symbol represents the current directory. Therefore, `grep -r 'ERROR' .` will search for the string `ERROR` in all files in the current directory and its subdirectories.

Scripting with Bash

Bash (**Bourne Again SHell**) is one specific shell implementation that has gained widespread popularity and is the default shell for many Linux distributions. Bash scripts can automate repetitive tasks, handle file and text manipulation, control job scheduling, and much more.

> **Note**
> While Bash is a specific shell, the term "shell" is more generic and encompasses other shell implementations.

A **Bash script** is a plain text file that contains a series of commands. These scripts can be used to automate entire workflows and complex processes that you'd otherwise have to perform command by command on the command line.

To create a Bash script, use a text editor to write your script, save it with any name, and give it the `.sh` extension. For example, you might name your script, `script.sh`. You can also use Vim like so:

```
localhost:~/pipelines# vi run_pipeline.py
```

Figure 6.6: Creating a Bash script

In *Figure 6.6*, we are creating a Bash script using `vi`, and then providing the filename `run_pipeline.py`. Once you hit *Enter*, you must hit the *i* key on the keyboard to begin editing the file. If you don't, you will not be able to edit it.

> **Note**
> For more information on using the `vi` Unix-based text editor, check out `https://www.redhat.com/sysadmin/get-started-vi-editor`, which goes deeper into this topic.

The first line of every Bash script should be `#!/bin/bash` (also known as a **shebang**) – this line tells the system that this is a Bash script and should be executed with the Bash shell.

Here's a simple Bash script:

```
#!/bin/bash

# This is a comment
echo "Hello, world!"
```

This script will simply print the string "Hello, world!" when it's run.

Once you're finished editing the text file, hit *Esc*, and type :wq to save and exit the editor. Then, hit *Enter*. You will be back to your most recent directory. To run a Bash script, use the bash command followed by the script name:

```
bash script.sh
```

Alternatively, you can make the script itself directly executable with the chmod command:

```
chmod +x script.sh
```

Then, you can run the script like this:

```
./script.sh
```

You can also use variables in your Bash scripts. Variables are declared using the $ symbol. It is important to not include space around the equal sign when assigning a variable to avoid errors. Here is a simple example:

```
#!/bin/bash

greeting="Hello, world!"
echo $greeting
```

In this script, greeting is a variable that stores the string "Hello, world!". The $ symbol is used to access the value of the variable.

Assessment

You have created a Bash script called script.sh in your current directory. However, when you try to run the script using ./script.sh, the terminal returns an error: "Permission denied". What command can you use to resolve this issue and why?

Answer

```
chmod +x script.sh
```

This issue arises because the script does not have the execute (x) permission. The chmod command is used to change the permissions of a file. The +x option adds execute permissions to the file. So, chmod +x script.sh will give execute permissions to script.sh, which will allow you to run the script with ./script.sh.

Assessment

In the context of Bash scripting, what does the line #!/bin/bash at the beginning of scripts signify and why is it important?

Answer

The line #!/bin/bash is known as the shebang. It is used to tell the system that the following script should be executed using Bash. This is important because different systems can have different default shells, and a script intended to be run with Bash might not work correctly if run with a different shell. By including #!/bin/bash at the start of your scripts, you ensure that they will be run using the correct interpreter regardless of the system's default shell.

Introducing control statements

Control statements, including conditional statements and loops, are an integral part of shell scripting, allowing you to incorporate decision-making and repetitive tasks in your scripts. As a data scientist, you might use control statements when automating data preprocessing, running different analyses based on certain conditions, or when building complex pipelines. This section will introduce the most commonly used control statements in Bash scripting.

Just like other programming languages, Bash provides conditional statements to control the flow of execution. The most common conditional statements in Bash are if, if-else, and if-elif-else.

Let's take a look at a simple if statement:

```
#!/bin/bash
x=10
if [ $x -gt 5 ]
then
   echo "x is greater than 5"
fi
```

In this script, if the value of x is greater than 5, the message x is greater than 5 is printed to the console.

As you can see, control statements are often paired with arithmetic operators. Here is a list of Bash arithmetic operators and their meaning:

Bash Arithmetic Operator	Meaning	Meaning Details
-lt	<	Less than
-gt	>	Greater than
-le	<=	Less than and equal to
-ge	>=	Greater than and equal to
-eq	==	Equal to
-ne	!=	Not equal to

Figure 6.7: Bash arithmetic operators

An if-else statement executes one block of code if the condition is true, and another block of code if it is false:

```
#!/bin/bash
x=10
if [ $x -gt 5 ]
then
  echo "x is greater than 5"
else
  echo "x is not greater than 5"
fi
```

In this case, if x is not greater than 5, the script prints x is not greater than 5.

Additionally, here is a good place to remind you that spacing is important to avoid errors when writing an if statement. Bash and Shell are less forgiving than Python when it comes to spacing and will produce an error if it is incorrect.

For multiple conditions, use if-elif-else:

```
#!/bin/bash
x=10
if [ $x -gt 10 ]
then
  echo "x is greater than 10"
elif [ $x -eq 10 ]
then
  echo "x is equal to 10"
else
```

```
    echo "x is less than 10"
fi
```

This script checks multiple conditions and executes different blocks of code depending on which condition is true.

Looping constructs, specifically `for` and `while` loops, are critical for executing tasks multiple times. Here's a `for` loop example:

```
#!/bin/bash
for i in {1..5}
do
   echo "This is iteration $i"
done
```

This script prints `This is iteration` x for each iteration from 1 to 5.

And here's a `while` loop example:

```
#!/bin/bash
x=1
while [ $x -le 5 ]
do
   echo "This is iteration $x"
   x=$(( $x + 1 ))
done
```

The script performs the same task as the previous `for` loop, but it uses a `while` loop that continues until x is greater than 5. The `x=$(( $x + 1 ))`, adds 1 to x during each iteration of the loop.

Understanding and employing these control statements in Bash scripts can help automate and streamline your data science workflows, making your operations more efficient and reproducible.

Assessment

Suppose you have an x variable and you want to write a script that prints `x is positive` if x is greater than 0, `x is negative` if x is less than 0, and `x is zero` if x is equal to 0. How would you construct this script using conditional statements?

Answer

You would use an `if-elif-else` statement. Here is an example of how you would construct the script:

```
#!/bin/bash
x=10
if [ $x -gt 0 ]
```

```
then
  echo "x is positive"
elif [ $x -lt 0 ]
then
  echo "x is negative"
else
  echo "x is zero"
fi
```

This script first checks whether x is greater than 0. If this condition is true, it prints x is positive. If it's false, it then checks whether x is less than 0. If this condition is true, it prints x is negative. If both conditions are false (i.e., x is not greater than 0 and not less than 0), it must mean x is equal to 0, so it prints x is zero.

Creating functions

Functions in Bash are blocks of reusable code that perform a certain action. They help structure scripts and avoid repetitive code, making scripts easier to maintain and debug. In data science, you might use Bash functions to perform recurring tasks such as loading data, processing files, or managing resources.

A function in Bash is declared with the following syntax:

```
function_name() {
  # Code here
}
```

function_name is the name of the function, which you'll use to call it. The code inside the curly braces { } is the body of the function.

Here's an example of a function that prints a greeting:

```
greet() {
  echo "Hello, $1"
}
```

This greet function prints "Hello" followed by the first argument passed to it. The $1 part is a special variable that refers to the first argument.

Once a function is defined, it can be called by its name. For example, to call the greet function, you would write the following:

```
greet "Data Scientist"
```

This line of code will print Hello, Data Scientist.

You can pass arguments to a function just like you would with a command. Inside the function, you refer to these arguments with $1, $2, and so on, where $1 is the first argument, $2 is the second, and so on. Here's a function that takes two arguments and prints them:

```
print_arguments() {
  echo "First argument: $1"
  echo "Second argument: $2"
}
```

To call this function with the arguments Data and Science, you would write the following:

```
print_arguments "Data" "Science"
```

This will print the following:

```
First argument: Data
Second argument: Science
```

In Bash, a function returns the exit status of the last command executed. You can explicitly specify a return status using the return statement, followed by an integer:

```
is_even() {
  if [ $(($1 % 2)) -eq 0 ]
  then
    return 0
  else
    return 1
  fi
}
```

This function checks whether the first argument is an even number. If it is, the function returns 0 (indicating success in Unix-like systems); otherwise, it returns 1.

Assessment

Imagine you are writing a Bash function that takes a filename as an argument and prints the number of lines in that file. What would that function look like?

Answer

The function might look something like this:

```
count_lines() {
  echo "The file $1 has $(wc -l < $1) lines"
}
```

In this function, `count_lines`, the `$1` argument is used to represent the filename passed to the function. The `wc -l < $1` command is used to count the lines in the file, and the entire `echo "The file $1 has $(wc -l < $1) lines"` command prints out a message with the filename and the line count.

Processing data and pipelines

As a data scientist, you often need to handle and process large datasets. Bash provides powerful tools for data processing and creating pipelines, which are sequences of processes chained by their standard streams. This allows the output of one command to be passed as input to the next. Several commands in Bash are incredibly useful for data processing. Here are a few examples:

- `cat`: Concatenates and displays the content of files.
- `cut`: Removes sections from lines of files.
- `sort`: Sorts lines in text files.
- `uniq`: Removes duplicate lines from a sorted file.
- `head filename` and `tail filename`: These commands output the first and last 10 lines of a file, respectively. You can specify the number of lines by adding `-n`, as in `head -n 20 filename`.

Here's an example of using `cat`, `sort`, and `uniq` to display the unique lines in a file:

```
cat filename | sort | uniq
```

The `cat` function displays the contents of the file. The pipe (|) takes the output of the `cat` function and sends it to the `sort` function, which sorts the line in the text. Then we use the pipe (|) function again to take the output of the `sort` function and send it to the `uniq` function. Finally, the `uniq` function removes any duplicate lines.

For more complex text-processing tasks, you might use commands such as `awk` and `sed`. Now, `awk` is a complete text-processing language that is ideal for data manipulation, while `sed` (stream editor) is a tool that parses and transforms text.

Here's an example of using `awk` to print the first column of a file:

```
awk '{print $1}' filename
```

In this command, `{print $1}` is an `awk` command that prints the first field (`$1`) of each line.

Meanwhile, `sed` is another tool useful for performing find-and-replace operations, substitutions, deletions, and more on text files or input streams. Here is an example of using `sed` to substitute the word `example` with the word `sample` in a text file:

```
sed 's/example/sample/' example.txt
```

Here's an explanation of what's happening:

- `sed` is the command for using `sed`
- `s/example/sample/` is the substitution pattern, where `s/` indicates a substitution, `example` is the search pattern to find, and `sample` is the replacement
- `example.txt` is the input file on which the substitution is being performed

Using pipes

Pipes are a powerful feature in Bash that allow you to create complex data processing pipelines. They allow you to stick multiple functions together like Lego blocks to make a complex pipeline.

Here's an example of a pipeline that processes a CSV file, removes the header, sorts the lines by the second column (assumed to be numeric), and writes the output to a new file:

```
tail -n +2 data.csv | sort -t, -k2,2n > sorted_data.csv
```

Here are the details of this pipeline:

- `tail -n +2 data.csv` outputs the content of `data.csv` starting from the second line (thus removing the header)
- `sort -t, -k2,2n` sorts the lines by the second column as a number
  - `-t,` specifies the comma as the field separator
  - `-k2,2n` specifies the second field as the sort key
  - `n` indicates that it should be sorted numerically
- `>` redirects the output to `sorted_data.csv`

Bash's data processing commands and pipelines provide powerful tools for manipulating and analyzing data. Learning how to use these features can make your work as a data scientist more efficient, especially when dealing with large datasets or complex data transformations.

Assessment

Imagine you have a CSV file with a header row. The file contains several columns of data, including a `Year` column. You want to sort the data by the `Year` column, which is the third column in the file. How would you accomplish this task in Bash?

Answer

You could use a combination of the `tail`, `sort`, and `>` commands to accomplish this task. The command would look something like this:

```
tail -n +2 filename.csv | sort -t, -k3,3n > sorted_filename.csv
```

The `tail -n +2 filename.csv` command removes the header row by printing all lines from `filename.csv` starting from the second line. The `sort -t, -k3,3n` command sorts the output by the third column (the `Year` column), treating the entries as numbers. The `-t,` option tells sort to use a comma as the field separator, and `-k3,3n` tells it to sort numerically on the third field. The `>` operator redirects the sorted output into `sorted_filename.csv`.

Using cron

cron is a powerful feature in Unix-like operating systems that allows users to schedule tasks (called cron jobs) to run automatically at specific times or on specific days. As a data scientist, you might use cron to automate tasks such as retrieving data, cleaning data, or running scripts at regular intervals.

The `crontab` (cron table) command allows you to create, edit, manage, and remove cron jobs. Here's an example of how you might use the `crontab` command to view your current cron jobs:

```
crontab -l
```

The `-l` option tells `crontab` to list the current user's cron jobs.

To edit your cron jobs, you would use the `-e` option:

```
crontab -e
```

This command opens the current user's `crontab` file in the default text editor. If no `crontab` file exists for the user, this command creates one.

A cron job is defined by a line in the `crontab` file, which consists of six fields:

```
*       *       *     *     *              command to be executed
-       -       -     -     -
|       |       |     |     |
|       |       |     |     +----- day of the week (0 - 6) (Sunday=0)
|       |       |     +------- month (1 - 12)
|       |       +--------- day of the month (1 - 31)
|       +----------- hour (0 - 23)
+------------- min (0 - 59)
```

Each field can be an asterisk (which means *any value*), a single value, a range of values, or a list of values or ranges separated by commas.

Here's an example of a cron job that runs a script every day at 2:30 PM:

```
30 14 * * * /home/user/data_script.sh
```

This line specifies that the `data_script.sh` script, located in `/home/user/`, should run at minute 30 of hour 14 (2:30 PM) every day.

By default, the output from a cron job is mailed to the owner of the `crontab` file. However, you can redirect the output to a file:

```
30 14 * * * /home/user/data_script.sh > /home/user/data_log.txt
```

In this example, the output of `data_script.sh` is redirected to `data_log.txt`.

Keep in mind that while cron is powerful and flexible, it also has some limitations and isn't the right tool for every job. However, there are tools, such as Airflow and Luigi, that make up for its shortcomings.

Assessment

You have a Python script called `data_update.py` that updates your data every week. The script is located in the `/home/data_scientist/` directory. How would you schedule a cron job to run this script every Monday at 1:30 AM?

Answer

To schedule this cron job, you would open your `crontab` file using the `crontab -e` command, and then add the following line:

```
30 1 * * 1 /usr/bin/python3 /home/data_scientist/data_update.py
```

This `cron` job is scheduled to run at minute 30 of hour 1 (1:30 AM) every Monday (1 in the day-of-the-week field). The command to run is `/usr/bin/python3 /home/data_scientist/data_update.py`, which executes the `data_update.py` script with Python 3. Please note that the path to Python might differ based on the specific system configuration.

Summary

In this chapter, we covered a broad range of topics related to basic shell and Bash scripting and command-Line operations for data scientists.

We began with an overview of navigating within the file structure and directory on a local computer or a virtual machine from the command line, explaining the use of basic commands for directory navigation. Then, we moved on to file and directory manipulation. In the subsequent sections, we delved into Bash scripting topics, discussing control statements and the use of Bash functions to create reusable pieces of code. We highlighted data processing and pipelines, demonstrating how to chain commands together to process text data. We also covered cron jobs for scheduling tasks and provided an overview of its syntax.

Gaining fluency in Bash scripts and basic shell commands will prepare you to engage with a variety of other CLI technologies commonly used in data science such as interfacing with the cloud providers (i.e.: AWS, Azure, GCP), Hadoop, Docker, Flask, or Kubernetes.

In our next chapter, we will look at version control with Git.

7

Using Git for Version Control

This chapter aims to prepare you for interview questions related to Git, a version control system integral to collaborative projects and data management.

Throughout these sections, you'll delve into the basics of creating and managing repositories and common Git operations, such as `config`, `status`, `push`, `pull`, `ignore`, `commit`, and `diff`. We will also highlight the common workflow patterns for a data scientist using Git and the crucial role of branches in this workflow.

The goal is to equip you with practical knowledge that you can leverage during your technical interviews, enabling you to demonstrate not only your data science acumen but also your adeptness at utilizing essential collaboration tools. Understanding these concepts is pivotal in today's data science landscape, as efficient version control and collaboration are as critical to a project's success as the scientific methods employed.

In this chapter, we will cover the following topics:

- Introducing repositories (repos)
- Creating a repository
- Detailing the Git workflow for data scientists
- Using Git tags for data science
- Understanding common operations

Introducing repositories (repos)

Repos are a version control system in a centralized storage location, holding all the files, directories, and version history of a project. A repository allows multiple developers to collaborate on a project, keeping track of changes made to the project's files over time, which is useful for projects with multiple data scientists and developers. It stores all the different versions of the files, along with metadata such as the author, timestamp, and description of each change.

There are many version control options that organizations might use. Some popular options include GitHub, BitBucket, GitLab, Azure DevOps repositories, and AWS CodeCommit.

It's important to note that there are multiple phases of version control. The major three are repos, a working directory, and a staging area. We've already explained what a repo is, but what are the other two?

A **working directory** is the directory on your local machine where you have cloned or initialized a Git repository. It contains all the project files that you can modify, create, or delete as part of your development process. When you make changes to files in the working directory, Git recognizes them as modifications to the project.

The **staging area** (aka **an index**) is an intermediary stage between your working directory and repo and is where the files of your project are ready to be tracked. Thus, it acts as a holding area for changes that you intend to include in the next available version of the project by taking a snapshot of the modified files. However, instead of committing to these changes directly from the working directory, you explicitly choose which changes to add to the staging area. In doing so, the staging area allows you to control which changes are included (committed), enabling you to selectively group related changes together or split them into separate commits.

Working with repos for version control is all about moving project files from one phase of the Git workflow to the next, which is demonstrated in *Figure 7.1*:

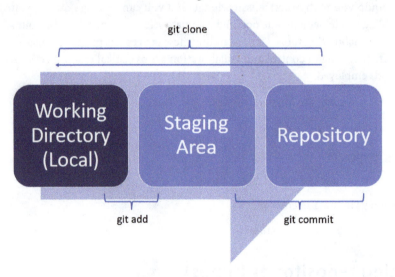

Figure 7.1: The Git workflow

Think of this concept as saving your progress in a video game. When you're actively playing, you are navigating your "working directory." However, if you want to save your progress, this is the equivalent of moving your progress to the "staging area." If you want to share your progress with friends, you might migrate your saved file to another console. This is the equivalent of a "repository."

Creating a repo

In this section, we'll cover the essential steps for creating a GitHub repository from an existing remote repository, as well as creating a local repository without an existing remote repository. Then, we will look at linking a local and remote repository. Let's begin!

Cloning an existing remote repository

When working as a part of a project team, a central repository has likely already been created. If you are working with a project that already exists, use the `clone` command to make a local copy of the repository. Cloning allows you to have a local copy of the project on your own computer, where you can work on it offline, experiment with it, and contribute your changes back to the project if you wish.

Here's how to clone a repository:

1. Retrieve a copy of the remote repository URL. If GitHub is your remote repository, then this can be found under the green **Code** button, currently on the **Code** tab of a project.

2. Open the terminal on your local machine.

3. Change the current working directory to the location where you want the cloned directory to be made.

4. Type `git clone`, and then paste the URL you copied earlier. If GitHub is your remote repository, then the command might look like this:

    ```
    git clone https://github.com/YOUR-USERNAME/YOUR-REPO-NAME.git
    ```

    You are passing the central remote repo URL as input to the `clone` command.

Afterward, Git will create a copy of the repo in your current directory.

Creating a local repository from scratch

When starting a new project from scratch, you can initialize a repo using `init` (meaning "initialize") inside of a local project folder. This will create a `.git` file on your machine; however, note that it is not visible by default on your computer's filing system (it will, however, show in your terminal):

```
git init <project-name>
```

This command will create a new repository in the current directory. Therefore, change to the directory where you want to make the repository first.

Once a repository has been created, it doesn't automatically start tracking your files. You need to tell Git which files to begin tracking by using the `add` command. This command places files in the staging area. Think of the staging area as an intermediate step between the working directory and the repository. It plays a crucial role in managing and organizing changes before they are committed to the repository.

Here is an example of using the `add` command:

```
git add <file_name>
```

Using the command in this method only adds one file to the staging area. However, you can use the `--all` option to stage all the files in the directory simultaneously:

```
git add --all
```

The example adds all the files in the directory to the staging area.

> **Note**
>
> If, for whatever reason, you need to reverse a staged add, use `git reset HEAD`, followed by a filename. This allows you to remove changes from the staging area without discarding the modifications in the working directory.

After adding a file to the staging area, you can use the `commit` command to move it to the repository. After executing this command, you are then asked to add a log message to your commit, which is basically a comment that describes your changes. You do this by adding the –m flag and then your message in parentheses. Here is an example of adding a message to a commit:

```
git commit -m "This is a message."
```

You want to be thoughtful with your message, since it will forever be part of the repo.

To summarize our recent discussion, here's how to create a new local repository:

1. Navigate to the directory where you want to create the repository.
2. Within this directory, initiate a new local repository with the `git init` command. You'll see output that says, **Initialized empty Git repository in [your directory]**.
3. If you want to copy an existing repo, use `git clone`.
4. Add files to your repository by creating new files or moving existing files into this directory.
5. After adding or modifying files, stage the changes by using `git add` command, which stages all changes in the directory and subdirectories. You may also use `git reset HEAD <file-name>` to reverse a staged file.
6. Commit these changes to your repository by using `git commit -m "Commit message"`, where `"Commit message"` is a message that describes the changes you've made.

At this point, you have a local repository with your initial project files.

Linking local and remote repositories

After creating a local repository, you can link it to a remote repository to easily share your code, collaborate with others, and have an online backup of your work.

Here's how to link your local repository to GitHub:

1. Create a new repository on GitHub (to avoid errors, do not initialize the new repository with README, `gitignore`, or `License` files; these can be added after your project has been pushed to GitHub).

2. Get the remote repository's HTTPS URL from the GitHub page (the same way as described in the instructions about cloning).

3. In the terminal, change the current working directory to your local project.

4. To add the URL for the remote repository where your local repository will be pushed, run the following command, replacing `https://github.com/YOUR-USERNAME/YOUR-REPO-NAME.git` with your repository's URL:

    ```
    git remote add origin https://github.com/YOUR-USERNAME/YOUR-
    REPO-NAME.git
    ```

5. Push the changes in your local repository to GitHub using `git push`:

    ```
    git push -u origin master
    ```

6. Now, your local repository is linked to your GitHub repository, and all your local changes can be pushed to the GitHub repository for safe-keeping and sharing.

7. To review the history of any project, use `git log` along with flags to learn details about the project through logs. Here are some examples:

    -   `git log -3 myfile.py` shows the last three commits to `myfile.py`
    -   `git log -since YYYY-MM-DD` shows the commits since the provided date
    -   `git log -author=<name>` shows all commits by the provided author

    There are many other flags that you can use. To look up other flags for any given command, use `git <command> -help`.

In summary, whether you're cloning an existing repository or creating a new one, you're setting up an environment where you can contribute to a project in a controlled and effective manner. Git and GitHub form the backbone of many modern data science workflows, and understanding these steps is crucial in preparing for a data science interview.

Assessment

Create a local copy of the `git@github.com:py-why/dowhy.git` repository in the `/home/project/code/` directory.

Answer

```
cd /home/project/code/
git clone git@github.com:py-why/dowhy.git
```

First, the `cd` command is used to change to the `/home/project/code/` directory. Now, once in that directory, the `clone` command is used to make a local copy of the repo within the directory.

Assessment

You've been working on a new data analysis project locally and want to share your progress with your colleagues via GitHub. Explain the process of creating a local repository and linking it to a remote repository on GitHub.

Answer

To create a local repository and link it to a remote repository on GitHub, follow these steps:

1. Create a new directory for your project on your local machine and navigate to it.

2. Initiate a new local repository within this directory with the `git init` command.

3. Add files to your repository by creating new files or moving existing ones into this directory.

4. Stage the changes with `git add --all` command, which stages all changes in the directory and subdirectories.

5. Commit these changes to your repository with `git commit -m "Commit message"`.

6. Create a new repository on GitHub. To avoid errors, don't initialize the new repository with "README", `.gitignore`, or "License" files.

7. Copy the remote repository's HTTPS URL from the GitHub page.

8. In the terminal, change the current working directory to your local project.

9. Add the URL for the remote repository where your local repository will be pushed with `git remote add origin https://github.com/YOUR-USERNAME/YOUR-REPO-NAME.git`, replacing the URL with your repository's URL.

10. Push the changes in your local repository to GitHub with `git push -u origin master`.

This process allows you to work on your project locally and then share your work via GitHub, making it available for others to see, clone, or contribute to.

Detailing the Git workflow for data scientists

Understanding Git workflows is a key competency for data scientists. As we've discussed before, Git allows you to track changes, revert to previous versions, and collaborate with others. In this section, we'll describe a typical Git workflow for a data scientist and explain the concept of a branch, an important feature in Git.

A **branch** in Git is essentially a unique set of code changes with a unique name. Each repository has one default branch (usually called `master` or `main`) and can have multiple other branches. The branches are used to develop features isolated from each other. When you want to create a new feature or experiment with something without disturbing the main line of development, you create a new branch. If the experiment is successful, you can merge these changes into the main branch. If it's unsuccessful, you can discard the branch, and it won't affect your main branch or repository.

Here is the typical Git workflow for a data scientist:

1. **Create a new branch for your task**: If you're about to start work on a new feature or a bug fix, it's good practice to create a new branch. This keeps your changes organized and separate from the main branch. The command to create a new branch is `git branch new-branch-name`. To switch to this branch, you use the `git checkout new-branch-name` command.

2. **Add changes to the new branch**: Once you're on the new branch, you can make changes to your files and stage them with `git add filename.ext`, or `git add --all` to stage all changes.

3. **Commit the changes to the branch**: After staging the changes, you commit them with a descriptive message using `git commit -m "Your commit message"`.

4. **Push your changes to the remote repository**: After committing your changes, you can push them to the remote repository with `git push origin new-branch-name`.

5. **Open a pull request**: On GitHub, you can open a `pull` request, which allows others to review and discuss your changes. If you're collaborating with a team, this step is crucial for code review and collaborative debugging.

6. **Merge your branch into the main branch**: After your changes have been approved, you can merge them into the main branch. On GitHub, this can be done with the merge button in your `pull` request. Locally, you would first check out to the main branch with `git checkout main` and then merge your branch with `git merge new-branch-name`.

7. **Pull the latest changes from the main branch**: Other people might have made changes to the main branch while you were working on your feature. To make sure your local main branch is up to date, use `git pull origin main`.

8. **Repeat the process for a new feature or bug fix**: After your changes are merged into the main branch, you can repeat the process, starting from *step 1*, for your next task.

It's important to note that these steps describe one possible workflow with Git, known as the **feature branch workflow**. Different teams and projects might use different workflows. In the context of data science, you might use branches to experiment with different models or data processing techniques.

For example, you might create a new branch to experiment with a new machine learning model. If the model improves your results, you can merge it back into the main branch. Here, assume that you are working with GitHub on a classification problem and you want to explore the results using the decision tree algorithm. At the end of the example, we delete the local copy of the branch we created because it is now merged into our main branch:

```
git branch decision-tree
git checkout decision-tree
…(assumes that you're updating your code files and review results)
git add --all
git commit -m "Explored results using decision tree algorithm"
git push origin decision-tree
…(assumes that a submitted a pull request and it was approved)
…(assumes the branch was merged into main in GitHub)
git checkout main
git pull origin main
git branch -d decision-tree
```

In a technical interview, you might be asked to describe how you would use Git in a collaborative project, or to describe a time when you used Git to manage different versions of a data science project. Understanding the concept of branches and the basic Git workflow can help you answer these questions confidently.

Assessment

You are working on a new feature for a data science project. Describe the series of Git commands you would use to create a new branch, add and commit your changes, and then push these changes to the remote repository.

Answer

First, you would create a new branch using `git branch new-branch-name`, and then switch to it with `git checkout new-branch-name`. Once you've made your changes, you would stage them for commit, using `git add filename.ext` for specific files or `git add --all` for all changes. After staging the changes, you'd commit them with a message, using `git commit -m "Your commit message"`. Finally, you would push the changes to the remote repository with `git push origin new-branch-name`.

Assessment

Explain the importance of using different branches when working on a data science project and how it might influence your workflow.

Answer

Using different branches is crucial in a data science project because it allows for experimentation without affecting the main line of development. For instance, if you want to test a new algorithm or dataset, you can create a new branch and make changes there. If the changes improve your project, you can merge them into the main branch. If they don't, you can simply discard the branch without it affecting your main code base. This ensures that the main branch only contains code that is tested and works properly. Moreover, in a collaborative environment, branches provide a way for multiple team members to work simultaneously on different features without conflict.

Using Git tags for data science

Tagging in Git is a way to mark specific points in your repository's history as being important. Typically, people use this functionality to mark release points (v1.0, v2.0, and so on). In this section, we'll cover the concept of tagging and how it can benefit data scientists.

Understanding Git tags

There are two types of tags that Git recognizes, lightweight and annotated. A **lightweight** tag is similar to a branch that doesn't change. It's just a pointer to a specific commit. **Annotated** tags, however, are stored as full objects in the Git database. Using the annotated tag is generally recommended because it is fully tracked and contains more info than the lightweight tag.

To create an annotated tag in Git, you can use the `git tag -a` command, followed by the tag name (usually the version), and then the message, such as the following:

```
git tag -a v1.0 -m "my version 1.0"
```

To view the tags in your repository, you can use the `git tag` command.

Using tagging as a data scientist

Tagging can be especially useful for data scientists for versioning models or experiments. For instance, if you have trained a machine learning model and want to keep track of its versions, you could use a tag to mark the `commit` that produced the model.

You could also use tags to mark the `commit` that generated a particular result or figure. This can be extremely helpful in ensuring the reproducibility of results, which is a crucial aspect of data science.

In addition, using tags can help data scientists collaborate more effectively. Team members can use tags to share the specific versions of the code they are working on, or to indicate which versions produce the best results.

In a technical interview, you might be asked about your strategies for managing versions of your code or ensuring reproducibility. Discussing your experience with Git tagging can help demonstrate your commitment to good practices in data science.

Remember, Git tagging is not a replacement for proper experiment tracking in data science, which should also record parameters, performance metrics, and other important details of each experiment. However, it can be a helpful tool to manage your code base and collaborate with others.

Understanding common operations

Understanding the basic commands of Git is paramount for anyone working in the field of data science. In the previous section, we delved into how to set up a GitHub repository, either by cloning an existing repository or starting a new one from scratch. In this section, we will explore common Git operations that will help you manage your repositories more effectively.

So, let's take a look at some operations:

- **Configuring Git** (`config`): Git's configuration settings can be found in the `.gitconfig` file, which is usually located in the user's home directory. To modify these settings, use the `git config` command. Set your name and email address, which will be attached to each commit you make:

```
git config --global user.name "Your Name"
git config --global user.email "youremail@domain.com"
```

Check your settings:

```
git config --list
```

- **Checking the status** (`status`): The `git status` command provides information about the current state of the repository, including untracked files, changes that are staged but not yet committed, and the branch you're currently on:

```
git status
```

- **Pushing changes** (`push`): The `git push` command allows you to send the commits from your local repository to a remote repository:

```
git push origin master   # Push changes to the master branch
```

If you want to share your tags with others, you need to use the `git push --tags` command:

```
git push origin --tags
```

- **Pulling changes** (`pull`): The `git pull` command is used to fetch and download content from a remote repository and immediately update the local repository to match that content:

  ```
  git pull origin master   # Pull changes from the master branch
  ```

- **Checking differences** (`diff`): The `git diff` command is used to display the differences between two points in your repository:

  ```
  git diff                 # Show differences not yet staged
  git diff --staged        # Show differences between staged
  changes and the last commit
  ```

- **Ignoring unnecessary files** (`.gitignore`): When working on a project, there are often files that you don't want Git to track, such as log files or files containing sensitive information. This can be managed with a `.gitignore` file in your repository's root directory. Patterns defined in this file will apply to all files in the repository. Here is an example of a `.gitignore` file:

  ```
  *.log
  *.csv
  secrets/*
  ```

  In this example, all `.log` and `.csv` files will be ignored, as well as all files in the `secrets/` directory.

These commands form the backbone of many interactions with Git and are crucial for efficient version control. As a data scientist, being comfortable with Git is a must, as it not only allows you to collaborate with other team members but also lets you keep track of changes, allowing you to revert back to previous versions when necessary.

In the context of a technical interview, a good understanding of Git indicates that you are familiar with the basic version control tooling used in data science and software development, which can make a strong impression on potential employers. Remember, learning Git is not just about memorizing commands but also understanding how these commands can be integrated into your workflow, improving productivity and collaboration.

Assessment

You are working on a data science project and have made several changes to your Python scripts. However, you realize that you've made a mistake and want to see what has changed since your last `commit`. Which Git command would you use, and what does it do?

Answer

You would use the `git diff` command. This command shows the differences between the changes you've made in your working directory and the last `commit`. It's used to review the changes you've made before staging and committing them, which is useful when you want to confirm your changes or when you're troubleshooting. The output shows the lines that have been added or removed.

The following code shows an example, where `a/file.txt` and `b/file.txt` are different versions of the same file:

```
diff --git a/file.txt b/file.txt
index ce01362..5d34e82 100644
--- a/file.txt
+++ b/file.txt
@@ -1 +1 @@
-I love coding
+I love to learn
```

Assessment

During your work on a machine learning project, you've accumulated several large `.csv` files containing intermediate results. These files are cluttering up your Git `status` and you don't want to accidentally `commit` them. How can you tell Git to ignore these files?

Answer

To tell Git to ignore certain files, you can use a `.gitignore` file. This file resides in the root directory of your repository. In this case, you would add `*.csv` to your `.gitignore` file, which tells Git to ignore all `.csv` files in the repository. This is very useful to exclude unnecessary files, such as temporary files, logs, or files with sensitive data, from being tracked by Git.

You should be careful to only ignore files that truly don't need to be in the repository, as ignoring important files could lead to lost work or inconsistencies between different versions of a project.

Summary

In this chapter, we explored the core fundamentals of Git, an essential tool for data scientists looking to effectively manage and collaborate on projects. We kicked things off by guiding you through setting up a GitHub repository. This involved the creation of a new repository, both from scratch and by cloning an existing remote repository. We provided a step-by-step walk-through, offering a straightforward approach to establishing and preparing your local repository for development work.

Following this, we navigated through the common Git operations that form the backbone of interaction with this tool. We explored essential commands such as `config`, `status`, `push`, `pull`, `ignore`, `commit`, and `diff`, laying out their functions and demonstrating their usage with practical examples. Additionally, we delved into the concept of branches, a critical feature of Git that allows you to segregate your changes and efficiently manage different project versions, using tags to highlight specific points in your repository. Finally, we described a typical Git workflow for a data scientist, providing a roadmap for creating, modifying, and merging branches in the context of a data science project.

With this knowledge, you are now equipped to handle version control and collaboration tasks effectively, a vital skill for any technical interview.

In our next chapter, we will look at analyzing data with statistics.

Part 3: Exploring Artificial Intelligence

The third part of this book covers various data mining techniques, how they work, the assumptions they make, their evaluation criteria, and their applications. We start with the foundations of inferential statistics, followed by increasingly more advanced data mining tasks, including the most popular machine learning models, neural networks, and generative AI. This part ends with helpful tips on deploying an effective MLOps strategy.

This part includes the following chapters:

- *Chapter 8, Mining Data with Probability and Statistics*
- *Chapter 9, Understanding Feature Engineering and Preparing Data for Modeling*
- *Chapter 10, Mastering Machine Learning Concepts*
- *Chapter 11, Building Networks with Deep Learning*
- *Chapter 12, Implementing Machine Learning Solutions with MLOps*

8
Mining Data with Probability and Statistics

In this chapter, you will be introduced to the vital world of statistics, which serves as the foundation of applied data science. An understanding of these concepts is crucial for drawing meaningful conclusions and making informed decisions and predictions from data. This knowledge is not just an intellectual exercise; it equips you with essential tools to excel in advanced data science interviews by allowing you to uncover hidden insights within datasets.

This chapter will guide you through the essential aspects of classical statistics, including the analysis of populations and samples, measures of central tendency and variability, and the intriguing realms of probability and conditional probability. You'll also explore probability distributions, the **central limit theorem** (**CLT**), experimental design, hypothesis testing, and confidence intervals. This chapter concludes with a focus on regression and correlation, giving you comprehensive tools to understand relationships within data and make confident predictions.

In this chapter, we will cover the following topics:

- Describing data with descriptive statistics
- Introducing populations and samples
- Understanding the CLT
- Shaping data with sampling distributions
- Testing hypotheses
- Understanding Type I and Type II errors

Describing data with descriptive statistics

Descriptive statistics are values that summarize the characteristics of a dataset. Before working on a project, data scientists use descriptive statistics to better understand the dataset they are working with. Think of it like exploring a treasure chest of information, with descriptive statistics as your guide to finding important details.

In your technical interview, you will be expected to be able to understand and use descriptive statistics. In this section, we will look at how to measure the central tendency of our dataset, then explore measures of variability or how dispersed and how much spread our dataset has.

Measuring central tendency

We are exposed to measures of centrality every day. For instance, if you live in the US, you might have heard that home prices in the state of California of the US are, on average, higher than in the state of Ohio. Of course, this doesn't mean that every home in California is more expensive than every home in Ohio, but if we could collect a lot of homes from each state in two separate baskets and draw from each one, more often than not, the home in the California basket will cost more than the home pulled from the Ohio basket. We know this because, on average, according to Redfin, the median price of California homes averages $798,600 [1], while those in Ohio average $249,400 [2].

Measures of **central tendency** provide a snapshot of a dataset's typical or central value, helping us to understand where data tends to cluster.

When discussing measuring centrality, we often use measures such as mean, median, and mode:

- The **mean** represents the arithmetic center of data and is calculated by summing up all the values in a dataset and dividing the sum by the number of observations. For example, the mean of [4, 6, 8, 10] is (4 + 6 + 8 + 10) / 4 = 7.

- The **median** is the middle value in a dataset when the observations are arranged in ascending or descending order (if the dataset has an odd number of observations, the median is the middle value itself, and if the dataset has an even number of observations, the median is the average of the two middle values). For instance, the median of [4, 6, 8, 10] is 7.

- The **mode** is the value or values that occur most frequently in a dataset. Unlike the mean and median, the mode does not rely on mathematical calculations. Sometimes, a dataset may have multiple modes (bimodal, trimodal, and so on), or there may be no mode if all values occur with the same frequency. For example, if given the values [4, 6, 8, 8, 10], the mode is 8.

How do we identify when we should use the mean over the median and vice versa? Let's consider an example where we try to estimate the average income of a population. Suppose we have a dataset of incomes for a specific population, and the distribution of income is highly skewed, with a few extremely high-income individuals. In this situation, using the mean as the measure of central tendency may not accurately represent the typical income of any singular person in the population. This is because high-income earners are outliers.

In these situations, you will want to use the median, which provides a measure of centrality that is not influenced by outliers as much as the mean. For example, suppose I take the average of my neighbor's annual income and average it with Jeff Bezos. In that case, the resulting product will look nothing like the average wage for any given individual in America – not unless I averaged it with many more people and with much smaller wages. Even then, the average would not be representative of the wage most people take home. Thus, the median is more valuable as it represents the middle value of the dataset when arranged in ascending or descending order.

Recall from *Chapter 4* that you can quickly visualize your numerical data as a histogram or box plot to see if the data is highly skewed or has significant outliers. Based on this insight, you can then decide if the average or median would better represent your data:

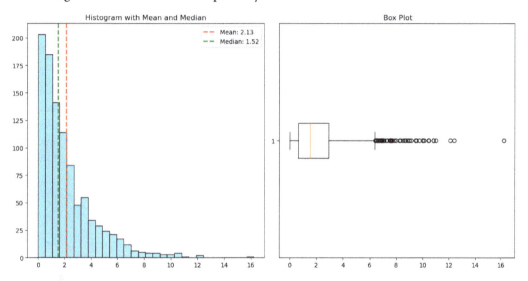

Figure 8.1: Illustration of a histogram (left) and box plot (right) of a skewed distribution

> **Note**
>
> If the income distribution follows a symmetrical, bell-shaped distribution (a normal distribution), the mean and median would likely be very close to each other. In such cases, using either the mean or the median as the measure of central tendency would provide a representative estimate of the average income.

Measuring variability

As you've seen thus far, mean, median, and mode are not enough to explain a dataset's shape. While measures of centrality measure the central tendency of data (that is, the tendency of the data's central statistics), variability helps us understand the spread or dispersion of data points.

Variability is the measure of a dataset's spread. For example, the mean wage in a country might be $54,000, but how much does this figure vary from person to person? Is this average the result of low variability (that is, everyone's wage is somewhat close to $54,000) or high variability (that is, there is a small minority that makes billions of dollars, but their wages are saturated by the vast majority who make under $30,000)?

In short, variability provides insights into how data points deviate from the central tendency. Three commonly used measures are as follows:

- **Range**: The range of a dataset is given by subtracting the smallest figure from the largest figure. For example, if a group of students in a class received {5, 12, 24, 9, 18} as values for a quiz, the range of the values is 19 or 24-5=19.

- **Interquartile Range (IQR):** The IQR is the range of the middle 50% of the data set. It is calculated by subtracting the first quartile (the 25th percentile) from the third quartile (the 75th percentile). The IQR is less sensitive to outliers than the range and is often used to summarize skewed data.

- **Standard deviation**: The standard deviation is the data's standardized distance (or deviation) from the mean of the dataset. It helps in understanding the variability of a process. The standard deviation is a more robust measure of variability than the range as it considers how every value in the dataset contributes to the dispersion. It has the same units as the original data, making it easier to interpret in context. Additionally, standard deviation is often represented by the Greek letter sigma, σ, while the square of standard deviation (σ^2) is variance.

Assessment

Suppose you are working with a dataset containing employees' salaries in a large organization. The CEO's salary is significantly higher than everyone else's, causing a skew in the salary distribution. Which measure of central tendency (mean, median, or mode) would be the most appropriate to represent a "typical" employee's salary, and why?

Answer

The median would be the most appropriate measure of central tendency to represent a "typical" employee's salary in this case. The reason is that the median is less affected by outliers or extreme values, such as the CEO's salary, compared to the mean. The mean takes into account all values, so an extremely high value can significantly skew it upwards. The median, on the other hand, is the middle value of the dataset when ordered from smallest to largest and thus can provide a more representative "typical" value when the data contains significant outliers.

Assessment

You are examining a dataset of exam scores for two classes of students who were taught the same course by two different teachers. Class A has a much smaller standard deviation in exam scores than Class B. What can you infer from this about the distribution of scores in each class, and what might this suggest about the two teaching methods?

Answer

A smaller standard deviation in exam scores for Class A implies that the scores in Class A are closer to the mean score and hence more consistent, with less variability. The scores for Class B, with a larger standard deviation, are more spread out from the mean, indicating greater variability.

In terms of teaching methods, while we can't make definitive conclusions from this data alone, it might suggest that the teaching method for Class A led to more consistent understanding among students, while the teaching method for Class B resulted in a wider range of understanding. It could also suggest that the teacher for Class A had a teaching style that was effective for a larger proportion of the students compared to the teacher for Class B. However, these are just hypotheses and would need further investigation and more information to support them since many other factors could be influencing the distribution of scores in each class.

Introducing populations and samples

Statistics is the art of extracting meaningful insights from data, and it all begins with a thorough understanding of populations and samples. In this section, we will explore the fundamental concepts that underpin statistical analysis by distinguishing between populations and samples.

Understanding these concepts is important because they form the basis for generalizing observations from a subset of data to a larger group. By investigating the intricacies of populations and samples, you will gain the necessary tools to make sound inferences and draw reliable conclusions from the data you encounter. So, let's embark on this enlightening journey and uncover the foundations of statistical analysis.

Defining populations and samples

In the realm of statistics, a **population** refers to the entire group of individuals, objects, or events that we are interested in studying. For instance, if we wanted to research the average height of all adults in a country, the population would comprise every adult within that country. It would not include other countries or children, for example.

However, studying an entire population is often impractical or impossible due to factors such as time, cost, or accessibility. This is where samples come into play. A **sample** is a subset of the population that we select to represent the larger group. By randomly selecting and analyzing a sample, we can draw meaningful conclusions about the population as a whole. In data science, we are almost always working on a dataset that represents the sample of a larger population:

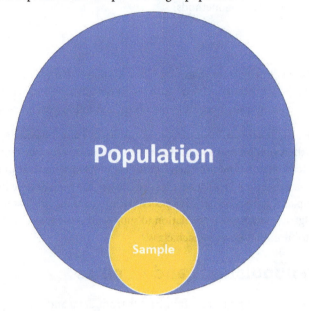

Figure 8.2: Illustration of a sample of a population

Representing samples

The key to reliable statistical analysis lies in the representativeness of the sample. A **representative sample** accurately reflects the characteristics and diversity of the population it is drawn from. Achieving representativeness requires carefully considering factors such as sampling methods, sample size, and potential biases. Simple **random sampling** is one of the most straightforward methods of sampling, where every individual in the population has an equal chance of being selected. This approach ensures that the sample is unbiased and, therefore, representative of the population, provided the sample size is sufficiently large.

Sampling bias occurs when we do not acquire a representative sample. It is a sample that is systematically skewed and does not accurately represent the population. For example, imagine that you are running for class president in your high school, and you want to conduct a poll to understand your chances of winning. The population of your poll is your high school, and the sample you collect is from all seniors. However, collecting a sample of just seniors creates a biased, unrepresentative sample of the high school. Perhaps you have more friends in the senior class because you are a senior.

You will be very disappointed on election day if first-year students, sophomores, and juniors overwhelmingly vote against you!

In data science, it is essential to be aware of various sources of bias, such as selection bias, non-response bias, and measurement bias, to minimize their impact on statistical analysis and predictions.

Now, let's suppose you conduct a sufficiently randomized sample of 100 students from all four classes at the school. After tallying the poll, it looks like 80% of them are willing to vote for you on election day – congratulations! But not so fast. Suppose you conducted another poll the following day, and it turns out that only 75% of that sample committed to voting for you. You collect more samples on different days, and the results are all different. What gives? You might be experiencing sampling bias across the different days. For example, if after day 1's poll, your poor Statistics class grade was released, your polling numbers may dip on day 2. Or, this could just be a case of sampling error.

Reducing the sampling error

The **sampling error**, also known as the standard error of a sample, is the natural variation that occurs between different samples from the same population. Even if the true proportion of students supporting you is consistent throughout the entire student body, each sample will capture a slightly different proportion due to just random chance. This makes sense, because rarely do we sample anything and get the same results every time. The sampling error reminds us that the estimates obtained from our samples are not exact replicas of the true population proportion and that uncertainty and variability are always at play in statistics.

To mitigate the impact of the sampling error, we can increase the sample size and number of samples. The standard error is calculated as the standard deviation of the population statistic divided by the square root of the sample size. Mathematically, it can be represented as follows:

$$SE = \frac{\sigma}{\sqrt{n}}$$

Here, we have the following:

- SE is the standard error

- σ is the standard deviation of the population statistic

- $\sqrt{n}$ is the sample size

As the sample size grows larger, the standard error decreases. Similarly, increasing the number of samples also reduces the sampling error. The more samples you collect, the better you can estimate the true population parameter by considering the range and distribution of estimates across the samples. To calculate the overall standard error when combining results from multiple samples, you can compute the standard deviation of the sample statistics across all the samples and divide it by the square root of the total number of samples. This accounts for the variability between the sample estimates.

Understanding the sampling error enables us to quantify the uncertainty in our estimates and make reliable inferences.

Assessment

You are studying the average commuting time of workers in a large city. Explain how you would define the population and a potential sample for this study. What are some of the considerations you would have to bear in mind when choosing the sample to ensure it's representative of the population?

Answer

In this case, the population is made up of commuting works who live in a large city. The sample might be a subset of these workers selected for the study, perhaps based on certain criteria such as accessibility or willingness to participate in the study. It's important to ensure that the sample is randomized and representative of the population as a whole, which means it should reflect the diversity of commuting times across different areas of the city, different professions and ages, and other factors that might affect commuting times. Potential biases, such as choosing more people who live in certain areas of the city, or more people from certain professions, should be carefully avoided to ensure that the sample is not skewed and can accurately represent the population.

Assessment

Describe the concept of the sampling error and how it affects the reliability of estimates drawn from a sample. What methods can be employed to reduce the impact of sampling error?

Answer

The sampling error is the natural variation that occurs between different samples from the same population. It signifies that the estimates obtained from individual samples are not exact replicas of the true population parameters; there is always some level of uncertainty and variability involved in statistics. This impacts the reliability of estimates because a high sampling error could lead to estimates that deviate significantly from the actual population parameters.

Reducing the impact of the sampling error can be achieved by increasing the sample size or the number of samples. Mathematically, the standard error is calculated as the standard deviation of the sample statistic divided by the square root of the sample size. Thus, as the sample size increases, the standard error decreases. Similarly, by increasing the number of samples collected, the overall standard error can be reduced as it allows for a better estimation of the true population parameter by considering the range and distribution of estimates across the samples.

Understanding the Central Limit Thereom (CLT)

Now that we've learned about sampling, now's the time to introduce one of the most important concepts in classical statistics – the **Central Limit Thereom (CLT)**.

The CLT

Measuring the center of data is not as simple as just calculating the mean, median, or mode. The CLT states that regardless of the original population distribution's shape, when we repeatedly take samples from that population and each sample is sufficiently large, the distribution of the sample means will approximate a normal distribution. This approximation becomes more accurate as the size of each sample becomes larger. This theorem plays a crucial role in measuring centrality by allowing us to make reliable estimates using these measures. In turn, the CLT enables us to estimate the population mean with greater accuracy, making the mean a powerful tool for summarizing data. It also indirectly influences the estimation of the median and mode. As the sample size increases, the distribution of individual observations becomes less skewed, enhancing the reliability of the median and mode as a measure of centrality.

The CLT also allows us to accept the **assumption of normality**, which allows us to rely on the normal distribution of sample means, even when the population distribution is not normal. Many statistical techniques and tests rely on the assumption of normality to ensure the validity of the inferences made. When the population follows a normal distribution, the CLT enables us to make accurate inferences about population parameters using sample means. This assumption allows us to use parametric tests.

> **Note**
>
> Many parametric hypothesis tests (such as t-tests and z-tests) rely on the assumption of normality to make valid inferences. These tests assume that the population from which the sample is drawn follows a normal distribution. The CLT comes into play by allowing us to approximate the distribution of the test statistic to a normal distribution, even when the population distribution is not strictly normal. This approximation enables us to perform these tests and make reliable conclusions.

Demonstrating the assumption of normality

In the previous section, we talked about the CLT and how it supports the assumption of normality. Let's look at a simple example to demonstrate how both work together.

Let's conduct an experiment where we roll a die repeatedly. We will be using a fair six-sided die, meaning that the die has not been altered, and there is an equal chance that when rolled, it might land on any of its six values. Since there is an equal chance of rolling any of the values on the die, this is considered a uniform distribution. In our experiment, we will repeatedly roll five times. Every time we roll the dice five times, we take the mean of our five die rolls. This is considered a sample. We repeat this process 10 times, computing 10 means:

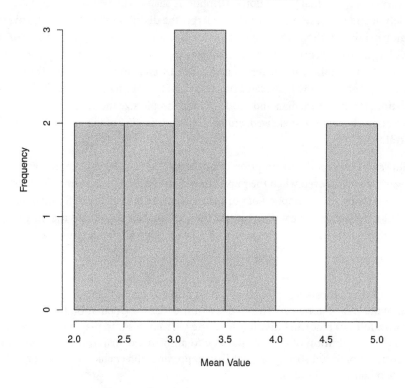

Figure 8.3: Distribution of dice roll samples – 10 times

Now, let's perform the same exercise, but this time, we'll replicate the experiment 100 times (resulting in 100 samples) instead of 10:

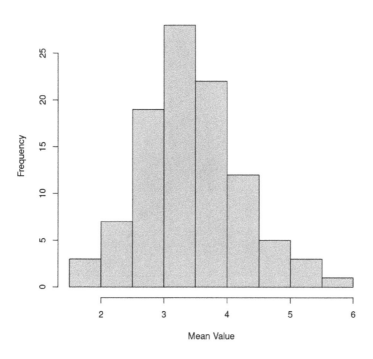

Figure 8.4: Distribution of dice roll samples – 100 times

Lastly, let's repeat the experiment one more time, only with 10,000 samples instead of 100:

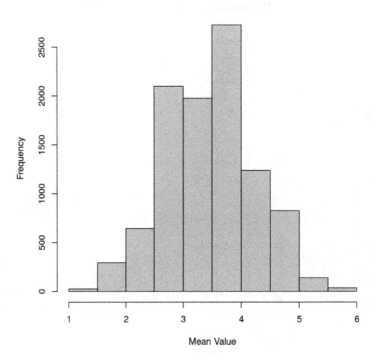

Figure 8.5: Distribution of dice roll samples – 10,000 times

Notice that the sample mean distribution now resembles a normal distribution, even though we know that rolling dice theoretically fits a uniform distribution. This illustrates the CLT – if you take a sufficiently large sample of random items from a population (typically 30 or more), regardless of the shape of the distribution of those items (as in our example of a uniform distribution from the die), the average of those samples will tend to approximate a normal distribution. This approximation becomes more accurate with larger sample sizes.

Assessment

Can you explain what the CLT states and why it is important in statistical analysis? How does it contribute to the measurement of centrality in a dataset?

Answer

The CLT is a fundamental theorem in statistics stating that, regardless of the population distribution's shape, when we repeatedly take sufficiently large samples from that population and calculate their means, the distribution of these sample means will approximate a normal distribution. This approximation becomes more accurate as the size of each sample increases. The CLT is crucial because it enables inferences about the population based on sample data, particularly regarding the population mean.

In terms of centrality, the CLT is primarily concerned with the mean. It asserts that with increasing sample sizes, the sample means tend to form a normal distribution, even if the original population distribution is not normal. This characteristic enhances the reliability and significance of the sample mean as a measure of central tendency, especially in making inferences about the population mean. However, the CLT does not directly impact the reliability of other centrality measures such as the median and mode, which depend on different aspects of the data distribution.

Assessment

Describe how the assumption of normality is linked with the CLT and how it influences the application of various statistical tests.

Answer

The assumption of normality in statistical analysis is closely linked to the CLT. According to the CLT, even if the population distribution is not normal, as the number of data points in each sample increases, the distribution of the sample means will approximate a normal distribution. This tendency toward a normal distribution in sample means is essential for the validity of many statistical tests, such as t-tests and z-tests, which are categorized as parametric tests.

Parametric tests typically rely on the assumption of a normally distributed population, particularly when working with small sample sizes. However, with larger samples (that is, samples containing more data points), the CLT becomes increasingly significant. In these cases, even if the population distribution is not normal, the CLT ensures that the distribution of sample means approaches normality. This approximation of normality in larger samples is crucial for the applicability of parametric tests, enabling the distribution of the test statistic to be treated as normally distributed. Consequently, this allows for reliable conclusions about the population parameters based on the sample data.

Shaping data with sampling distributions

If you've ever taken an introductory statistics course, you were probably taught that theoretical distributions (such as the ones we will discuss in this section) are a way to describe the central tendency and variability of a given numeric variable. Depending on the situation, it's often more appropriate to use one distribution over the other. Although this is an accurate summary of probability distributions, it's important to understand why we use them, and how you should think about them in a data science context (instead of that of a social sciences context, which is often how traditional introductory statistics classes are taught).

Probability distributions

Probability distributions are fundamental concepts in statistics and probability theory that describe the likelihood of various outcomes in a random experiment or process. In the world of data science, these distributions play a crucial role in modeling and understanding uncertainty. By studying the properties and characteristics of different probability distributions, we can gain insights into real-world phenomena, make predictions, and perform statistical inference. In this section, we will explore the major probability distributions that are commonly used in statistics and data analysis. Each distribution will be introduced, followed by a detailed explanation of its characteristics, formulas, and example scenarios where it is applicable.

To begin understanding probability distributions from a data scientist context, I want you to regard them as "shapes of data." As a data scientist, you will be leveraging countless datasets of various content and sizes. Discrete and continuous numeric variables in a dataset can be represented using probability distributions. A **discrete variable** is one where the values are real numbers that do not have partial values (for example, counts of items, proportions, ratios, or fractions). **Continuous variables** are numeric values that can hold any value between negative infinity and positive infinity. Given the distribution of the variable, you can make some useful assumptions about it, such as how to calculate probabilities associated with the dataset, and which models can be applied to the dataset given the confirmation of its assumptions.

Uniform distribution

The **uniform distribution** represents outcomes where each value within a given range is equally likely. We talked about this distribution briefly in the previous section, where we were running experiments rolling a die. In that case, the probability that the die would fall on one through six was equally likely to happen or a one out of six (1/6) probability. Another example of uniform distribution includes randomly selecting a card from a deck of cards. When selecting one random card from a 52-card deck, the probability for any card is one out of 52 (1/52).

In the context of data science, uniform distribution is frequently used in simulations and bootstrapping methods. It's also the foundational building block for generating random numbers in algorithms and models. This distribution is usually very easy to understand and explain. However, it may be too simplistic for complex real-world phenomena. While it's suitable for situations with equal likelihoods, it may not capture the nuances of datasets with more intricate structures.

Normal and student's t-distributions

The **normal distribution**, also known as the Gaussian or Z distribution, is perhaps the most widely used and essential probability distribution. It is characterized by its bell-shaped curve and is completely determined by its mean (μ) and standard deviation (σ). The **z-score** is a standardized value that measures how many standard deviations a given data point is from the mean. It allows us to convert any value from a normal distribution to the corresponding value on the standard normal distribution, making it a useful tool for probability calculations. Here is the z-score formula:

$$Zscore = \frac{(x - \mu)}{\sigma}$$

Here, we have the following:

- x is the data value

- μ is the mean of the normal distribution

- σ is the standard deviation

Let's consider an example of adult male heights. The heights of adult males in a given population often follow a normal distribution. Suppose the mean height is 175 cm and the standard deviation is 6 cm. Using the normal distribution, we can calculate the probability of finding a male with a height between 170 cm and 180 cm.

The **t-distribution** is the normal distribution's "cousin." The biggest difference is that it's generally shorter and has fatter tails. It is used instead of the normal distribution when the sample sizes are small. In t-distributions, the values are more likely to fall further from the mean. One thing to note is that as the sample size increases, the t-distribution converges to the normal distribution.

The binomial distribution

The **binomial distribution** models the number of successes in a fixed number of independent Bernoulli trials. A Bernoulli trial is a random experiment with two possible outcomes. In simpler terms, it describes the outcomes of repeated experiments where there are only two possible outcomes, often referred to as "success" and "failure."

A scientist might use the binomial distribution when computing the probability of flipping a fair coin 10 times and getting exactly four heads and six tails. In this scenario, we repeated the experiment of flipping a coin 10 times, and in each of these flips, the probability of receiving a head or tail was the same in each instance.

As a data scientist, it is important to remember that when using the binomial distribution, the probability of each success must also be the same for each trial. Also, there can only be two possible outcomes (hence "bi") for each of the trials. Finally, you cannot use a binomial distribution if the trials are not independent. For example, if a person is repeatedly selecting one card from a deck of cards but not returning the card to the deck each time, you cannot use the binomial distribution to model the probability of them selecting three spades over 10 tries. The chance of them selecting a spade card changes each time they draw a card because they are not returning it to the deck.

Here are a couple of more examples of when a data scientist might use the binomial distribution:

- **Modeling binary outcomes**: When dealing with experiments or processes that have exactly two possible outcomes (for example, pass/fail, on/off, and yes/no), the binomial distribution can be a perfect model.

- **Quality control and manufacturing**: In industries where the quality of products is critical, data scientists can use the binomial distribution to model the number of defective items in a batch. This aids in process optimization and quality assurance.

- **Marketing campaign analysis**: Data scientists can apply the binomial distribution to evaluate the success of marketing campaigns by analyzing the number of conversions (successes) versus non-conversions (failures) among targeted customers.

- **Healthcare research**: In medical trials, the binomial distribution can be used to model the number of patients responding positively to a treatment versus those who do not.

- **Sports analytics**: In sports, analyzing the number of wins and losses in a series of games can be modeled using the binomial distribution.

- **Election forecasting**: Predicting election outcomes based on sampled voter intentions, where voters can choose between two candidates, can also be represented with a binomial distribution.

The Poisson distribution

The **Poisson distribution** is the probability of a given number of (discrete) independent events happening in a fixed interval of time, and is commonly used in queuing theory, which answers questions like "How many customers are likely to purchase tickets within the first hour of announcing a concert?" These events must occur with a known constant mean rate (λ) and are independent of the time since the last event.

When using this distribution, a data scientist must remember the following aspects:

- Each event must be independent of the others.

- These must be discrete events, meaning that events occur one at a time.

- It is assumed that the average rate of occurrences, λ, is constant over the time interval. In the ticket purchasing example given earlier, it is assumed that the rate of ticket purchases will remain the same over the first hour and it doesn't suddenly increase in the last 10 minutes of the hour. A data scientist should validate these model assumptions when looking to use the Poisson distribution.

You know that data fits a Poisson distribution if the variable of interest is discrete and independent, and if it answers the question of how many events happen per a regular interval of time. Here are a few more scenarios when you should think about using the Poisson distribution:

- **Call center modeling**: A data scientist can model the number of calls a call center receives in an hour based on historical data, assuming a constant average rate

- **Website traffic analysis**: Analyzing the number of hits or visits to a web page within specific time intervals can be modeled using the Poisson distribution

- **Natural events**: Studying the number of earthquakes in a particular region over a year or the number of meteorites of a certain size hitting the Earth in a century are examples of Poisson processes

- **Service systems**: The number of customers arriving at a bank or a gas station in a fixed period can be modeled with the Poisson distribution

- **Healthcare**: In medicine, a Poisson distribution might be used to model the number of occurrences of particular incidents, such as the number of births in a hospital in a day

- **Quality control**: In manufacturing, it might describe the number of defects found in a particular sample of items

Exponential distribution

Similar to the Poisson distribution, the **exponential distribution** is a continuous distribution that simply models the interval of time between two events. You can also think of this as the probability of time between Poisson events. The exponential distribution models the time between consecutive events in a Poisson process, where events occur at a constant average rate (λ). It is often used to model waiting times and lifetimes of certain processes. For example, the time between consecutive visits to a website follows an exponential distribution with an average rate of 0.1 visits per minute. We can calculate the probability that a visitor will arrive within the next 10 minutes.

This distribution assumes that the events occur at a constant rate and that each event is independent of each other. A data scientist would want to check these assumptions are reasonable before modeling a process with this distribution.

Here are additional examples of where a data scientist might use an exponential distribution:

- **Lifetime modeling**: The exponential distribution is used to model the lifetime of products, machinery, and electronic components, representing the time until the first failure

- **Service systems**: It can describe the time between consecutive arrivals of customers in a system, such as a bank or a call center

- **Natural phenomena**: The time between occurrences of certain types of natural events, such as earthquakes or meteor showers, can be modeled with the exponential distribution

- **Medical research**: It can be used to model the time between successive occurrences of an event, such as the intervals between heartbeats or the time until the onset of a specific disease

Geometric distribution

The **geometric distribution** models the number of independent Bernoulli trials needed before observing the first success. For example, in basketball, if a player has a 70% chance of making a free throw (p=0.7), we can use the geometric distribution to calculate the probability of the player making the first free throw on their second attempt. Similar to the binomial distribution, we assume that each trial has two possible outcomes (success or failure), is independent of the others, and that the probability of success is the same for each trial. However, remember that the binomial distribution looks to model the number of successes over a fixed number of trials, while the geometric distribution models the number of trials required to achieve the first successful trial.

Again, here are some examples of where a data scientist might use a geometric distribution:

- **Reliability analysis**: The geometric distribution can model the number of uses of a product until it fails for the first time. This might be applied in industrial contexts to understand product longevity.

- **Marketing campaigns**: In marketing, this distribution might be used to model the number of contacts required to make the first sale to a new customer.

- **Medical trials**: In healthcare, it could represent the number of trials needed to achieve the first successful treatment in a series of independent treatments.

- **Ecology**: In environmental studies, it might describe the number of species sampled until the first endangered species is found.

- **Quality control in manufacturing**: The geometric distribution can model the number of items inspected until the first defective item is found.

The Weibull distribution

The **Weibull distribution** is a versatile distribution that's used in reliability engineering and survival analysis. It can model various shapes, including exponential (special case) and bathtub curves. Without getting too heavy into the math, the Weibull distribution is useful because of its flexibility, which is afforded to this distribution by two parameters: scale (λ) and shape (k). More specifically, the Weibull distributions are often used to model the time until a given technical device fails, but it also has other applications.

Here are some examples of when a data scientist might use the Weibull distribution:

- **Survival analysis**: In medical research, it's often used to model the time until the occurrence of certain events, such as the time until death in a population of patients with a specific disease

- **Weather forecasting**: It can be used to model wind speeds to help with designing wind turbines or predicting storm damages

- **Economics and Finance**: Some economic and financial phenomena that do not follow the normal distribution may be modeled using the Weibull distribution

- **Quality control in manufacturing**: It can model various aspects of the manufacturing process, such as the time until the first failure of a product:

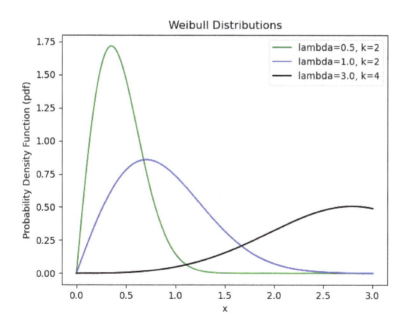

Figure 8.6: Three different forms of the Weibull distribution

Assessment

What are probability distributions and how are they utilized in the context of data science?

Answer

Probability distributions are fundamental concepts in statistics and probability theory that describe the likelihood of various outcomes in a random experiment or process. In the context of data science, these distributions play a critical role in modeling and understanding uncertainty. By studying the properties and characteristics of different probability distributions, data scientists gain insights into real-world phenomena, make predictions, and perform statistical inference. Given a certain distribution of a variable in a dataset, useful assumptions can be made, such as how to calculate probabilities associated with the dataset, and which models can be applied given the confirmation of the distribution's assumptions.

Assessment

Can you describe some of the major probability distributions used in statistics and data science, such as uniform distribution, normal distribution, t-distribution, binomial distribution, Poisson distribution, exponential distribution, geometric distribution, and Weibull distribution?

Answer

Here are the definitions for the different distributions:

- **Uniform distribution**: This represents outcomes where each value within a given range is equally likely. It's simple yet essential for describing uniformly random events.

- **Normal distribution**: Also known as the Gaussian or Z distribution, it is perhaps the most widely used distribution. It's characterized by a bell-shaped curve and is determined by its mean and standard deviation. A related concept is the z-score, which measures how many standard deviations a given data point is from the mean.

- **T-distribution**: It is similar to the normal distribution but has shorter, fatter tails. It's used when sample sizes are small. As the sample size increases, the t-distribution converges to the normal distribution.

- **Binomial distribution**: It models the number of successes in a fixed number of independent Bernoulli trials. A Bernoulli trial is a random experiment with two possible outcomes: success or failure.

- **Poisson distribution**: It represents the probability of a given number of independent events happening in a fixed interval of time. It's commonly used in queuing theory and related applications.

- **Exponential distribution**: This distribution models the time between consecutive events in a Poisson process, where events occur at a constant average rate. It's often used to model waiting times and lifetimes of certain processes.

- **Geometric distribution**: This models the number of independent Bernoulli trials needed before observing the first success. It can answer questions such as "How many trials until the first success?"

- **Weibull distribution**: This is a versatile distribution that's used in reliability engineering and survival analysis. It can model various shapes, including exponential, and is often used to model time until a given technical device fails, among other applications. Its flexibility is afforded by two parameters: scale and shape.

Testing hypotheses

In this section, we will review **hypothesis testing**, which is a statistical method that's used to make inferences about population parameters based on sample data. It involves formulating two competing hypotheses – the null hypothesis (H_0) and the alternative hypothesis (H_a) – and then using sample data to determine which hypothesis is more likely to be true.

The **null hypothesis**, or what I like to call "business as usual," is the default assumption or status quo for any given scenario. It's also often considered the "least interesting" scenario. For example, if I want to test whether or not changing my sneakers makes me a better runner, the sneakers not affecting my running abilities is the null hypothesis since there is no significant difference, effect, or relationship between the variables. Oftentimes, researchers are interested in rejecting the null hypothesis.

The **alternative hypothesis** is the opposite of the null hypothesis (mutually exclusive) as it represents the claim (that is, the hypothesis) being tested. It suggests that there is a significant difference, effect, or relationship in the population, given the contents of the sample.

Although the computations behind identifying critical values against given experiment parameters are beyond the scope of this book, we will go over the basics of what each statistical test does, and in what situations you may use them. There are many programs, including Python, R, and other statistical programs, that can run these tests. The hypothesis testing procedure involves the following steps:

1. Formulate the null hypothesis and the alternative hypothesis.

2. Randomly sample the population and calculate the appropriate test statistic (for example, t-statistic, z-score, or chi-squared statistic) from the sample.

3. Determine the appropriate probability distribution for the test statistic under the assumption that the null hypothesis is true.

4. Find the **p-value**, which is the probability of observing a test statistic as extreme as the one obtained, assuming the null hypothesis is true (the p-value measures the strength of the evidence against the null hypothesis).

5. Compare the p-value to a pre-determined significance level (alpha) to make a decision. In the data science industry, it is common to use a significance level of 5%. Therefore, if your p-value is below 5%, we reject the null hypothesis. If it is greater than the 5% threshold, we fail to reject the null hypothesis.

Note

We will primarily focus on the most common forms of parametric hypothesis testing since non-parametric testing is beyond the scope of this chapter.

Understanding one-sample t-tests

A **one-sample t-test** is a statistical procedure that compares the mean of a sample to a predetermined value to determine whether the observed difference is statistically significant or if it likely occurred due to chance alone.

For example, suppose we want to verify that the average male sea otter population in the Pacific Northwest of the US is maintaining a healthy weight, which we hypothesize to be 75 lbs. Since it's impractical to measure the entire population, we collect a sample of 50 male sea otters. Then, we calculate the sample mean and standard deviation. These values are used to compute a t-statistic, which will help us determine if the population mean is statistically significantly different from 75 lbs.

Understanding two-sample t-tests

A **two-sample t-test** (that is, two samples assuming equal variance test) determines whether there is a statistically significant difference between the means in two unrelated groups. For example, consider comparing the mean number of hours spent emailing per week by married respondents (population 1) and single respondents (population 2). The "Email Hours Per Week" variable is the test variable.

The **independent samples t-test** examines whether the difference between the mean number of hours married respondents spent emailing, and the mean number of hours single respondents spent emailing, is significantly different. To do this, we take samples from each population and compare their distributions. Are they significantly different? When in doubt, it's best to use an independent-sample t-test. This is appropriate for "between-subjects" designs where two groups of subjects are intended to differ on a critical manipulation.

Now, suppose we want to investigate whether there is a significant difference in the test scores of two study groups, Group A and Group B. Each group consists of different students, and the two groups were taught using different teaching methods:

- **Group A test scores**: [78, 86, 88, 92, 75, 82, 80, 85, 89, 94]
- **Group B test scores**: [72, 79, 84, 90, 81, 76, 88, 80, 83, 91, 85, 87]

We want to determine if there is a statistically significant difference in their average test scores. The hypothesis statements are as follows:

- **Null hypothesis**: There is no significant difference between the average test scores of Group A and Group B ($\mu A - \mu B = 0$)

- **Alternative hypothesis**: There is a significant difference between the average test scores of Group A and Group B ($\mu A - \mu B \neq 0$)

Understanding paired sample t-tests

In your journey as a data scientist, statistical testing will be a cornerstone of your work, often used to verify hypotheses and draw conclusions from the data you've collected or analyzed. One statistical technique that you'll likely encounter frequently, particularly when dealing with related samples, is the paired sample t-test, also known as the dependent sample t-test.

A **paired sample t-test** is a statistical procedure that determines whether the mean difference between two sets of observations is zero. The two sets of observations are typically dependent on each other – for example, the same set of individuals measured at two different time points or under two different conditions.

This test is applicable when you have two quantitative measurements, and these measurements are paired or related in some way. The "pairing" refers to the fact that each data point in one dataset is uniquely linked to a data point in the other dataset. In other words, there's a one-to-one correspondence between the values in the two sets. These scenarios can often be seen in the following areas:

- **Before-and-after observations**: Here, the same individuals, items, or events are measured before and after a treatment or intervention – for instance, measuring student test scores before and after an educational program

- **Matched pairs**: The pairs of observations come from two distinct groups, but each pair is matched or related in some way, such as twins, paired geographical locations, or matched units

Once you identify that you have paired data, the paired sample t-test can be used to compare the means of the two samples. The test assumes a null hypothesis that the true mean difference between the paired samples is zero and an alternative hypothesis that it is not. Depending on the test's result, you'll either reject or fail to reject the null hypothesis. Keep in mind that the paired t-test assumes that the differences between pairs follow a roughly normal distribution.

Understanding ANOVA and MANOVA

Analysis of Variance (**ANOVA**) and **Multivariate Analysis of Variance** (**MANOVA**) are powerful statistical tests that are often utilized by data scientists to analyze the differences among group means and their associated procedures. They offer an extension of the two-sample t-test to scenarios with more than two groups or variables.

ANOVA

ANOVA compares the means of three or more independent groups to test if they are significantly different from each other. The "business as usual" null hypothesis (H_0) posits that all group means are equal. The alternative hypothesis (H_a) is that at least one group mean is different.

We can represent this symbolically:

- H_0: $\mu1 = \mu2 = \mu3 = ... = \mu n$ (where μi represents the mean of each group)
- H_a: At least one μi is different

ANOVA is most appropriate when the following conditions are met:

- **Dependent variable**: The dependent variable is continuous (interval/ratio)
- **Independent variable**: The independent variable is categorical with at least three levels (different groups or categories)
- **Assumptions**: The data should meet the assumptions of independence, normality, and homogeneity of variance

MANOVA

MANOVA, an extension of ANOVA, is used when there are two or more dependent variables. The null hypothesis (H_0) claims that the multiple population mean vectors of the different groups are equal. The alternative hypothesis (H_a) asserts that they are different.

MANOVA is most suitable when the following conditions are met:

- **Dependent variables**: There are two or more continuous dependent variables
- **Independent variable**: The independent variable is categorical with at least three levels
- **Assumptions**: The data should meet the assumptions of multivariate normality and homogeneity of variance-covariance matrices

Chi-squared test

The **chi-squared test** is a non-parametric statistical method that's frequently employed in data science, particularly when dealing with categorical data. It is useful for assessing the relationship between two categorical variables in a sample.

There are two types of chi-squared tests:

- **Chi-squared test of independence**: Assesses if there is a significant association between two categorical variables
- **Chi-squared goodness of fit test**: Determines if the observed frequencies for a categorical variable match an expected set of frequencies

Let's delve a bit deeper into each of these tests.

Chi-squared test of independence

The chi-squared test of independence tests the null hypothesis (H_0) to see whether the two categorical variables are independent, with no association or relationship between them. The alternative hypothesis (H_a)asserts that there is an association or relationship between the two variables.

We can represent this symbolically:

- H_0: The variables are independent
- H_a: The variables are not independent

The chi-squared test of independence is appropriate under the following conditions:

- **Variables**: Both variables are categorical (nominal).
- **Observations**: Observations are independent, meaning each participant contributes only to one cell within the chi-squared table.
- **Assumption**: The assumption of a sufficiently large sample size is met. Generally, all expected frequencies should be at least 5.

Chi-squared goodness of fit test

The chi-squared goodness of fit test evaluates the null hypothesis (H_0) to see whether the observed frequency distribution of a categorical variable matches an expected frequency distribution. The alternative hypothesis (H_a) states that the observed distribution does not fit the expected distribution.

We can represent this symbolically:

- H_0: Observed frequencies = expected frequencies
- H_a: Observed frequencies $\neq$ expected frequencies

This test is suitable when the following conditions are met:

- **Variable**: The variable under consideration is categorical
- **Observations**: Observations are independent
- **Assumption**: All expected frequencies are at least 5

A/B tests

In the realm of data science, especially in fields such as marketing, product development, and user experience design, A/B testing, also known as split testing or bucket testing, is an essential method for comparing two versions of a single variable to identify which performs better.

An **A/B test** randomly assigns subjects to one of two groups: the control group (A), which receives the "business as usual" version, and the experimental group (B), which gets the version with the modification. The performance of the two groups is then compared to see if the modification resulted in any statistically significant improvement.

The hypotheses in an A/B test relate to whether a difference exists between the two versions:

- **Null hypothesis** (H_0): The null hypothesis posits that there is no difference in outcome between version A and version B
- **Alternative hypothesis** (H_a): The alternative hypothesis asserts that there is a difference in outcome between version A and version B

For example, if pA and pB represent the probability of a customer purchase for versions A and B of a website, respectively, then our null and alternative hypothesis would look like this:

- H_0: pA = pB
- H_a: pA ≠ pB

Applicability of A/B testing

A/B testing is most applicable in scenarios where you are testing a single modification between two versions. Some of the common conditions are as follows:

- **Controlled experiment**: You can control and randomly assign subjects to groups A or B
- **Single-variable testing**: You're testing a single change (for example, different headlines, page layouts, and color schemes)
- **Clear metrics**: There is a clear metric to measure success (for example, click-through rate, time spent on a page, purchase made, and so on)

Implementing A/B testing

The process of conducting an A/B test is as follows:

1. **Identify the variable**: Determine the element you want to test (for example, the color of a button, the length of a sales email, and so on).

2. **Formulate hypotheses**: Establish the null and alternative hypotheses.

3. **Split the sample**: Randomly assign your subjects to two groups, A (control) and B (treatment).

4. **Collect and analyze data**: Record the performance metric for each group, then compare the results to see if there is a statistically significant difference.

5. **Statistical test**: Perform a statistical test (such as a two-sample t-test) to check the significance of the difference.

6. **Make a decision**: If the p-value from your statistical test is less than your pre-set significance level (usually 0.05), reject the null hypothesis, concluding that your modification made a significant difference.

Assessment

Suppose you are working as a data scientist at a tech company that is developing a new feature for its main application. The company wants to determine if this feature will increase user engagement time. Describe how you would use hypothesis testing to help answer this question, and what specific tests might you use.

Answer

Hypothesis testing is an excellent method to answer such a question. First, I would define the null hypothesis to be that the new feature does not affect user engagement time, meaning that the average engagement time remains the same with or without the new feature. The alternative hypothesis would then state that the new feature does change user engagement time.

To test these hypotheses, I would suggest performing an A/B test, where users are randomly assigned to two groups: the control group (A), who use the application without the new feature, and the experimental group (B), who use the application with the new feature. The engagement times of both groups are then collected and compared.

Specifically, a two-sample t-test could be used to determine if there's a significant difference in the means of user engagement time between the two groups. If the p-value of the test is less than a pre-set significance level (usually 0.05), we would reject the null hypothesis in favor of the alternative hypothesis, indicating that the new feature has a statistically significant impact on user engagement time.

Assessment

You conducted a survey to understand whether customers prefer product A or product B. You hypothesize that there is a difference in preference. Explain which statistical test you would use to analyze the collected data and state the null and alternative hypothesis for this scenario.

Answer

The appropriate test to use in this scenario is the chi-squared test of independence. This test is used to determine whether there is a significant association between two categorical variables. Here, the two variables are the product (A or B) and preference (yes or no).

The null hypothesis for this test would be that there is no association between the product and preference, meaning that the product does not influence the preference. The alternative hypothesis would be that there is an association between the product and preference, meaning that the preference depends on the product.

We would collect data on customer preferences for both products and perform the chi-squared test of independence. If the resulting p-value is less than our chosen significance level (commonly 0.05), we reject the null hypothesis and conclude that there is a significant association between product and preference, which supports our original hypothesis that there is a difference in preference for the products.

Understanding Type I and Type II errors

In hypothesis testing, there is always a chance of making errors:

- A **Type I** error occurs when we reject the null hypothesis when it is true (this is also known as a false positive)

- A **Type II** error occurs when we fail to reject the null hypothesis when it is false (this is also known as a false negative):

	Null Hyp. True	Null Hyp. False
Reject Null Hyp.	Type I Error	Correct Rejection
Fail to Reject Null Hyp.	Correct Decision	Type II Error

Figure 8.7: Type I error vs. Type II

Understanding the nuances and implications of Type I and Type II errors is fundamental to hypothesis testing. In *Figure 8.7*, we see that Type I Error occurs at the intersection of the null hypothesis being true, and the action of rejecting the null hypothesis. This is similar to a pregnancy test coming back positive when the woman is not in fact pregnant (also known as a false positive result).

Simiarly, Type II Errors occur when the null hypothesis is false, but incorrectly fails to reject the null hypothesis. This is like having a pregnancy test that tells a pregnant woman that she is not pregnant (also known as a false negative).

Type I error (false positive)

A Type I error, or false positive, happens when we incorrectly reject a true null hypothesis. In simpler terms, it's an error of overreaction. We mistakenly believe there is a significant effect or difference when, in fact, there isn't. The probability of committing a Type I error is denoted by the Greek letter alpha (α), which corresponds to the significance level set for the test. If α is set to 0.05, for instance, we are willing to accept a 5% chance of committing a Type I error.

Type II error (false negative)

Conversely, a Type II error, or false negative, occurs when we fail to reject a false null hypothesis. This is an error of underreaction. We mistakenly believe there is no significant effect or difference when, in reality, there is. The probability of committing a Type II error is represented by the Greek letter beta (β).

One minus beta, or ($1-\beta$), gives us the power of the test, which is the probability of correctly rejecting a false null hypothesis. Hence, increasing the power of a test decreases the chances of committing a Type II error.

Striking a balance

The probabilities of committing Type I and Type II errors are inversely related. Reducing the risk of a Type I error (by choosing a smaller α) increases the risk of a Type II error, and vice versa. The key is finding the right balance between these two risks, and this balance depends on the context of the test and the potential implications of each type of error.

For example, in a medical context, a Type I error might lead to unnecessary treatment (false positive), while a Type II error might lead to a lack of treatment when it's needed (false negative). The relative costs and implications of these errors would guide the choice of α and, indirectly, the risk of a Type II error.

In conclusion, while you can never completely eliminate the risk of committing Type I and Type II errors, understanding these concepts, carefully choosing your significance level, and increasing your sample size (where possible) can help you manage and minimize these risks.

Assessment

In the context of a legal trial, where the null hypothesis is that the defendant is innocent (not guilty), can you explain what a Type I and a Type II error would correspond to and which one is considered more severe in this context?

Answer

In the context of a legal trial, a Type I error (false positive) would correspond to convicting an innocent person – that is, rejecting the null hypothesis (the defendant is innocent) when it is true. A Type II error (false negative) would correspond to acquitting a guilty person – that is, failing to reject the null hypothesis when it is false.

Generally, in legal contexts, a Type I error is considered more severe as it's based on the principle that "it is better that 10 guilty people escape than one innocent suffers." This is reflected in the idea of "innocent until proven guilty" and the requirement for proof "beyond a reasonable doubt." However, both types of errors are undesirable, and the legal system strives to minimize both.

Assessment

Describe how the chosen significance level (alpha, α) can impact Type I and Type II errors in a hypothesis test. What trade-offs might you have to consider when choosing the significance level?

Answer

The significance level, denoted by alpha (α), is the probability of rejecting the null hypothesis when it is true – that is, it directly corresponds to the probability of making a Type I error. If you set a lower significance level, say 0.01 instead of 0.05, you are decreasing the chances of making a Type I error; you are making the test more conservative and requiring stronger evidence to reject the null hypothesis.

However, making the test more stringent to avoid Type I errors increases the chances of making a Type II error – where we fail to reject a false null hypothesis. This is because you are setting a higher bar for the evidence required to reject the null hypothesis, which might lead to failing to reject the null hypothesis when it is false.

The choice of the significance level involves a trade-off between these two types of errors and will depend on which error has more severe consequences in the given context. For instance, in medical testing, a Type I error could lead to unnecessary treatment (possibly with side effects), while a Type II error could lead to missed treatment for a sick person. The relative costs and consequences of these errors guide the selection of the appropriate significance level.

Summary

In this chapter, we dove into the core fundamentals of data mining with statistics, which are often assessed during data science interviews. We reviewed the basics of probability, how to describe data using different measures of centrality and variability, how to estimate variables with population sampling, the relevance of the CLT and the assumption of normality, and reviewed probability distributions and hypothesis testing. By learning about these principles, you will be able to identify and describe relevant data statistics and make testable hypotheses. You will also avoid being fooled by misused statistics that manipulate our understanding of data.

Be aware that some interviewers will ask theoretical questions while others will want you to work out the solution to a problem. In either case, statistics is the backbone of many machine learning algorithms and experimentation designs, which are prominent in data science in all industries.

In the next chapter, we will build on our understanding of classical statistics by diving into pre-modeling concepts.

References

- [1] *California Housing Market*, from *Redfin* (June 2023): `https://www.redfin.com/state/California/housing-market`
- [2] *Ohio Housing Market*, from *Redfin* (June 2023): `https://www.redfin.com/state/Ohio/housing-market`

9

Understanding Feature Engineering and Preparing Data for Modeling

Wow – look how far you've come! Congratulations on making it to *Chapter 9*, where we will prepare you for machine learning concepts in the next chapter!

In this chapter, we will delve into the critical phase of pre-modeling. Here, you'll combine your knowledge of Python, data wrangling, and statistics.

While numerous data science texts emphasize the latest machine learning models, data preparation is the true foundation of successful prediction. This chapter is a vital bridge between collecting data and applying advanced machine learning techniques, emphasizing the data science principle, "garbage in, garbage out." Poor input data will yield unreliable results no matter how advanced a model is.

Pre-modeling data preparation is about ensuring our data is accurate, consistent, and relevant. Mastering this stage means understanding issues such as outliers, feature engineering, and imbalances. By addressing these, we will enhance the analysis quality, paving the way for robust and accurate predictive models.

This chapter covers a wide array of essential topics and techniques that data scientists commonly employ to prepare their data for modeling. Here's a brief overview of what you can expect to learn:

- Understanding feature engineering
- Applying data transformations
- Engineering categorical data and other features
- Performing feature selection
- Working with imbalanced data
- Reducing dimensionality

Understanding feature engineering

Feature engineering is a transformative process in data science that holds the key to unlocking the full potential of machine learning algorithms. As data scientists, we are tasked with analyzing the raw data and crafting new and informative representations of that data. Feature engineering involves selecting, transforming, and creating features that best capture the underlying patterns and relationships within data. By delving deep into the domain knowledge and leveraging our creativity, we can engineer features that amplify the predictive power of our models, improve accuracy, and enable better generalization of new data.

This section looks at the art and science of feature engineering, exploring a myriad of techniques and methodologies to extract meaningful insights from data and empower our machine learning algorithms to make informed and intelligent decisions.

> **Note**
>
> In this section, we will use Pandas for our feature engineering process. We covered some of Pandas' functions in *Chapter 3*.

Avoiding data leakage

Before discussing common data transformations and preprocessing techniques, we first need to acknowledge the importance of building reproducible and well-documented ML pipelines for maintaining data processing and modeling integrity. A major benefit of strong ML pipelines is that they ensure the modeling process avoids data leakage.

Data leakage is a phenomenon that leads to unreliable model performance due to the "leakage" of information beyond the training dataset that's being used during the creation of the model. This additional information can allow the model to learn something that it otherwise would not know (aka "peeking") and, in turn, invalidate the estimated performance of the model being constructed.

This is a mistake made by many novice data scientists who apply data transformations and preprocessing to the entire dataset prior to splitting the training set from the test set. This may lead to high bias and overly optimistic model performance.

To avoid data leakage:

- Split the dataset into training and testing sets.
- Train the transformations on the training data only, then use the results on the test set.

Here is an example of how to properly avoid data leakage with a normalization task:

1. Split the dataset:

    ```
    X_train, X_test, y_train, y_test = train_test_split(X, y, test_
    size=0.2, random_state=42)
    ```

2. Create the data transformation task pipeline:

    ```
    pipeline = Pipeline([ ('scaler', StandardScaler()), ])
    ```

3. Fit and transform the task to the training set:

    ```
    X_train_transformed = pipeline.fit_transform(X_train, y_train)
    ```

4. Transform the testing set with the training set task pipeline:

    ```
    X_test_transformed = pipeline.transform(X_test)
    ```

Using this technique, you can avoid data leakage and unreliable modeling results.

Handling missing data

Handling missing data is a common task in data preprocessing before applying machine learning algorithms. Missing data can introduce biases, errors, and instability in the analysis, leading to incomplete or misleading results. Moreover, some algorithms cannot handle missing data directly, making proper imputation essential for effective data processing. By replacing missing data, we maximize the utilization of available information and preserve underlying data patterns and relationships. This ensures an algorithm can operate without restrictions and enables accurate predictions and reliable outcomes.

Addressing different missing data mechanisms is equally important to avoid potential biases. Ultimately, replacing missing data enhances a dataset's accuracy, integrity, and usability, making it an indispensable part of data preparation for robust and trustworthy machine-learning applications.

Missing data can disrupt the accuracy of our analyses and models. Before proceeding with data imputation, it's crucial to identify the missing values within a dataset. Consider the following example:

```
import pandas as pd
# Sample dataset with missing values
data = {
    'A': [1, 2, None, 4, 5],
    'B': [6, None, 8, 9, 10],
    'C': [11, 12, 13, 14, None]
}
df = pd.DataFrame(data)
# Check for missing values in the dataset
print(df.isnull().sum()) #Output: A: 0, B: 1, C: 1
```

The code snippet uses the `isnull()` method to check for missing values in the DataFrame. The `sum()` function is then used to count the number of missing values in each column. We can see that columns B and C are missing one data point.

Now that we know our dataset is missing data, let's review how to handle this. As previously mentioned, there are a variety of methods to handle missing data, depending on the missingness pattern. **Missingness patterns** in data preparation refer to the systematic tendencies or structures in which data is absent, indicating the reasons or mechanisms behind the missing values. As promised, we will review a few examples of how to handle the following different scenarios:

- **Missing Completely at Random (MCAR)**: In this scenario, the missingness occurs randomly and is unrelated to any other variable in the dataset. One common approach to handling MCAR is to simply remove the rows with missing values:

```python
import pandas as pd
# Sample dataset with missing values (MCAR)
data = {
    'A': [1, 2, None, 4, 5],
    'B': [6, None, 8, 9, 10],
    'C': [11, 12, 13, 14, None]
}
df = pd.DataFrame(data)
# Removing rows with missing values (MCAR)
cleaned_df = df.dropna()
```

This script uses the `dropna()` function to remove any row from the dataset that has a missing value.

- **Missing Not at Random (MNAR)**: In MNAR, the missingness is related to unobserved or unrecorded values that are not random and may be related to the value itself. A common technique to handle MNAR is to use imputation methods to fill in the missing values, based on other available information:

```python
# Sample dataset with missing values (MNAR)
data = {
    'Age': [25, 30, None, 40, 45],
    'Income': [50000, None, 75000, 90000, None]
}
df = pd.DataFrame(data)
# Impute missing values with the mean of the 'Age' column
df['Age'].fillna(df['Age'].mean(), inplace=True)

# Impute missing values with the mean of the 'Income' column
df['Income'].fillna(df['Income'].mean(), inplace=True)
```

In this example, we use the `fillna()` function in combination with the `mean()` function to select the `Income` column mean and fill in any missing values within the column.

- **Missing at Random (MAR)**: In MAR, the missingness is systematic but depends only on the observed variables. One popular method to handle MAR is to use conditional imputation, where the imputed value depends on the values of other variables:

```python
# Sample dataset with missing values (MAR)
data = {
    'Gender': ['Male', None, None, 'Male', 'Female'],
    'Income': ['80-100k', '100-120k', '80-100k', '80-100k',
'100-120k']
}
df = pd.DataFrame(data)
# Impute 'Gender' based on the mode of 'Gender' for the
corresponding 'Income' value
mode_by_income = df.groupby('Income')['Gender'].apply(lambda x:
x.mode().iloc[0])
df['Gender'].fillna(df['Income'].map(mode_by_income),
inplace=True)
```

This script uses the `groupby()`, `mode()`, and `apply()` functions to find the mode (i.e., most common) gender for the different income categories. From there, it fills in any missing rows within the gender column with the most common gender.

Remember that the choice of handling missing data depends on the nature of the missingness, the dataset, and the goals of the analysis. Always consider the potential impact of the imputation on the overall analysis and modeling results.

Scaling data

Normalizing/scaling are preprocessing techniques that transform the features of the data into a consistent and comparable range, enabling algorithms to work more efficiently and producing accurate and reliable results. Two of the most commonly used normalizing/scaling techniques include min-max scaling and z-score scaling.

Min-max scaling is a technique that scales the data (typically the inputs) to a fixed range, typically [0, 1]. It transforms the data in such a way that the minimum value of the feature becomes 0 and the maximum value becomes 1.

Min-max scaling is expressed using the following formula for each data point, X, in a feature:

$$X_{new} = \frac{X - X_{min}}{X_{max} - X_{min}}$$

Here is how to implement this formula in Python:

```python
X_min_max = (X - X_min) / (X_max - X_min)
```

Here, `X_min` is the minimum value in the feature and `X_max` is the maximum value in the feature. The result is a min-max scaled feature that falls within the [0, 1] range.

Min-max scaling is particularly useful for the following:

- **Handling distance-based algorithms**: In distance-based machine learning algorithms (which we will cover in *Chapter 10*), such as *k*-means clustering or hierarchical clustering, the outcome is sensitive to the scale of the features. Min-max scaling ensures that each feature contributes equally to the distance calculations.

- **Distance-based algorithms (feature influence)**: When using distance-based machine learning algorithms such as *k*-nearest neighbors, hierarchical clustering, or when applying principal component analysis, you use algorithms that are sensitive to feature magnitudes/distance. Min-max-scaling the data helps ensure that each feature contributes equally to the distance calculations. This is important when the distance between data points is a significant factor in the algorithm.

Now, let's look at the other common technique for transforming our dataset features into a consistent and comparable range.

Z-score scaling is a technique that transforms data (again, typically the inputs) to have a mean of 0 and a standard deviation of 1. It centers the data around the mean and scales it relative to the spread of the data (the standard deviation).

The formula for scaling each data point, *X*, in a feature is given by:

$$X_{new} = \frac{X - \mu}{\sigma}$$

Here is how to implement it in Python:

```
X_standardized = (X - mean) / standard_deviation
```

Here, `mean` is the mean of the feature, and `standard_deviation` is the standard deviation of the feature. The result is a standardized feature with a mean of 0 and a standard deviation of 1. After scaling, the data tends to range from -3 to 3. However, it can be more or less, depending on the distribution of the data before scaling.

Z-score scaling is particularly useful for the following:

- **Feature influence**: Some machine learning algorithms can be significantly influenced by the scale and range of the input features, and *z*-scaling helps this issue. For example, you might have a dataset where you want to predict someone's BMI by measuring correlating features, such as their calorie intake (e.g., 1700 calories), age (e.g., 50 years), steps taken a day (e.g., 5,000 steps) or their blood sugar levels (e.g., 140 mg/dL). These variables are on entirely different scales. To ensure that one feature doesn't over-influence the model's performance, we use *z*-scaling, to ensure all features have the same relative influence by placing them on similar scales. Any relatively large or small values will now truly represent legitimate variation.

- **Dealing with outliers**: Scaling techniques such as Z-score scaling are less affected by outliers. Min-max scaling, conversely, can be influenced by extreme values and may not handle outliers as effectively.

So, when do you use min-max scaling over z-score scaling, or vice versa? The choice depends on the specific characteristics of the dataset and the requirements of the machine learning algorithm being used. Both techniques serve the purpose of transforming data to a comparable range, but their implications on the data may vary, making it important to consider the context and the nature of the data at hand. If unsure, experimenting with both min-max and z-score scaling and evaluating the model's performance can help determine the most effective preprocessing method.

Applying data transformations

Data transformations are vital steps in the data preparation journey. It ensures that data is prepped for data models with unique assumptions. This is achieved by transforming data from its current shape (or distribution) to another.In other words, transforming data from the empirical distribution to theoretical distributions.

In some cases, we need to transform our input variables to ensure that they're interpretable by the machine learning algorithm. An **input** variable (also known as a **feature**) is the columns of data, which typically explain some attribute of the data. In other cases, machine learning models require your **output** (aka a **response**) variable to have a certain distribution. An output variable is the column that we are trying to predict.

It certainly would be nice if the world accommodated our needs, but real-world data comes in all varieties! To remedy this scenario, you may have to perform a data transformation. In this section, we will explore common data transformation techniques.

Introducing data transformations

In the previous section, we discussed some popular techniques for transforming predominantly input data to adjust the scales or ranges. This section will discuss additional methods used to adjust the skew or relationships of data, including the response variable.

Remember your high school algebra course where you first learned about basic functions. If you recall, they looked something like this:

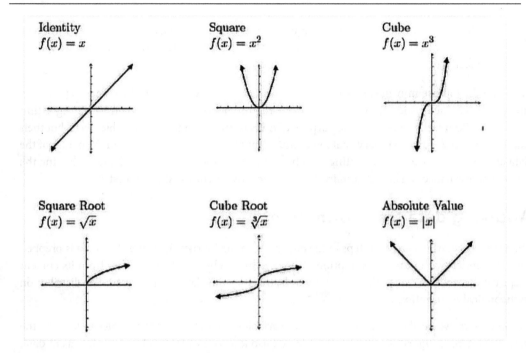

Figure 9.1: Base algebraic functions

Performing these transformations were a matter of applying the function of ($f(x)$) on x. Now imagine that these graphs are instead vectors of data. Imagine a graph's x and y coordinates represented by the x (the input variable/features) and y (the output variable/response) values of a dataset. In this scenario, each record is a vector.

When we talk about data transformation, we talk about transforming data vectors from one form to another, much like how we changed linear functions to square functions (also known as parabolas) by squaring each value. This process is helpful in pre-modeling for two primary reasons:

- To obtain the shape required by a model's assumptions before using the model on the data

- To revert the model's predictions to their original form (prior to the transformation)

Now that you understand the benefits of data transformations, it is important to know that there are many different types of data transformations. Here, we will summarize the following data transformation techniques:

- Log transformations

- Box-cox transformations

- Power transformations

- Exponential transformations

Let's get started.

Logarithm transformations

The **logarithmic (log) transformation** is beneficial when dealing with (typically right/positive) skewed data, where extreme values cause a long tail in the distribution. By taking the logarithm of the data, we can compress the range of high values and spread out the lower values, making the distribution more symmetric. For example, consider this example, where we have sales data:

```
import numpy as np
import pandas as pd
import matplotlib.pyplot as plt

# Create a left-skewed dataset
np.random.seed(42)
sales_data = np.random.exponential(scale=100, size=1000)

# Apply logarithmic transformation
log_transformed_data = np.log(sales_data)

# Plot the original and transformed data distributions
plt.figure(figsize=(10, 5))

plt.subplot(1, 2, 1)
plt.title('Original Data (Left-Skewed)')
plt.hist(sales_data, bins=30, edgecolor='black')
plt.xlabel('Values')
plt.ylabel('Frequency')

plt.subplot(1, 2, 2)
plt.title('Logarithmic Transformation')
plt.hist(log_transformed_data, bins=30, edgecolor='black')
plt.xlabel('Log-Transformed Values')
plt.ylabel('Frequency')

plt.tight_layout()
plt.show()
```

Here is what the `sales_data` variable looked like before and after the transformation:

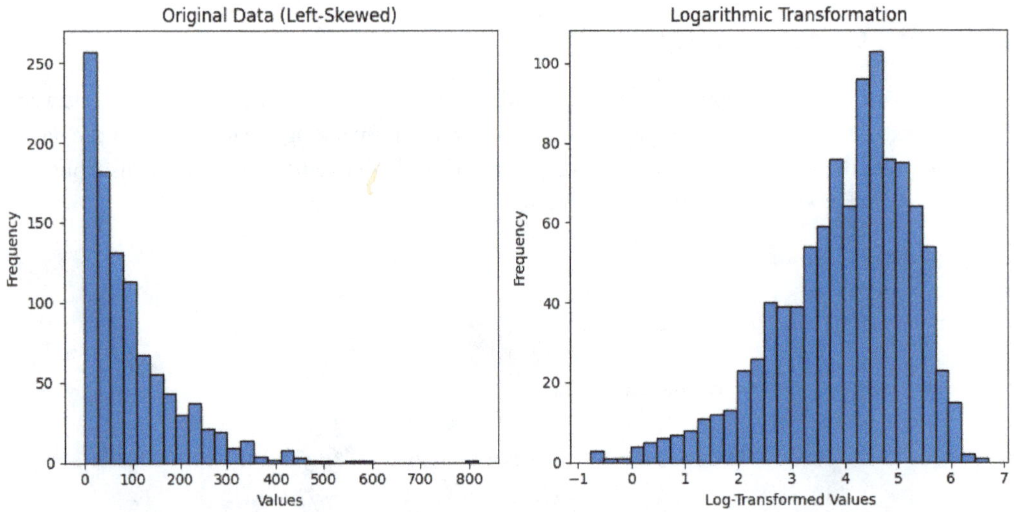

Figure 9.2: Sales data before and after a log transformation

The data distribution on the left does not follow a normal distribution with a noticeable bell-shaped pattern. Therefore, if you were attempting to use this data in a statistical test, such as those discussed in *Chapter 8*, you would be limited in the type of tests you could use. This is because many statistical tests, such as the one sample T-test, assume that your data comes from a normal distribution, and using the test on non-normal data can invalidate the results. However, the log transformation can convert our data closer to a normal distribution. That is what we see in the plot on the right. It is closer to a normal curve than the data plot on the left. We can still test whether the data in the plot on the right fits a normal curve, but assuming that it does, we now have more tests available for us to use.

> **Note**
> The logarithm transformation does not play well with negative values!

Power transformations

A **power transformation** is a family of data transformation techniques that involve raising each data point to a power (exponent). Different power values result in different transformations, allowing flexibility in shaping the distribution. Common power transformations include square root transformation (power = 0.5), cube root transformation (power = 1/3), and reciprocal transformation (power = -1).

Power transformations are valuable for handling data of various shapes. They are techniques used to adjust data with nonlinear relationships or inconsistent patterns. A key use of power transformations is to address **heteroscedasticity**, where the data's variability is uneven across its range. These transformations stabilize the variance in the data, making it more uniform and symmetrical. This is particularly beneficial for linear modeling. In the following figure, we can see this transformation process.

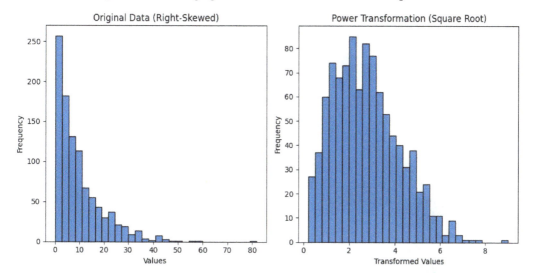

Figure 9.3: Distribution before and after a power transformation

The distribution on the left starts with a skewed dataset that is missing that familiar bell-shaped curve of a normal distribution. We apply the square root function, `np.sqrt()`, from the NumPy package to perform the power transformation on a variable. The histogram on the right displays the transformed data after applying the square root transformation. As a result of the transformation, the data becomes less skewed, and the distribution moves closer to a normal distribution.

Box-Cox transformations

A **Box-Cox transformation** is a family of power transformations that are designed to stabilize variance in our dataset and make it more closely follow a normal distribution. The Box-Cox equation can be seen here:

$$y\left(\lambda\right) = \begin{cases} y^\lambda, \ \lambda \neq 0 \\ Ln(y), \ x = 0 \end{cases}$$

The transformation is driven by an exponent, lambda (λ), which varies from -5 to 5. The Box-Cox family of transformations also includes both the logarithmic ($\lambda=0$) and square root ($\lambda=0.5$) transformations as special cases. It can automatically determine the best power parameter to stabilize variance and normalize data. It is often used to transform model features to fit a normal distribution in order to avoid heteroskedasticity, which occurs when the variance of data changes across different levels of the independent variable.

To implement this transformation in Python, we use `boxcox()` from the `scipy.stats` package:

```python
import numpy as np
import pandas as pd
import matplotlib.pyplot as plt
from scipy.stats import boxcox

# Create a right-skewed dataset
np.random.seed(42)
original_data = np.random.exponential(scale=100, size=1000)

# Apply Box-Cox transformation
transformed_data, lambda_value = boxcox(original_data)

# Plot the original and transformed data distributions
plt.figure(figsize=(10, 5))

plt.subplot(1, 2, 1)
plt.title('Original Data (Right-Skewed)')
plt.hist(original_data, bins=30, edgecolor='black')
plt.xlabel('Values')
plt.ylabel('Frequency')

plt.subplot(1, 2, 2)
plt.title('Box-Cox Transformation')
plt.hist(transformed_data, bins=30, edgecolor='black')
plt.xlabel('Transformed Values')
plt.ylabel('Frequency')

plt.tight_layout()
plt.show()
```

Here, we create a right-skewed dataset and apply the Box-Cox transformation, which can be seen in the following figure:

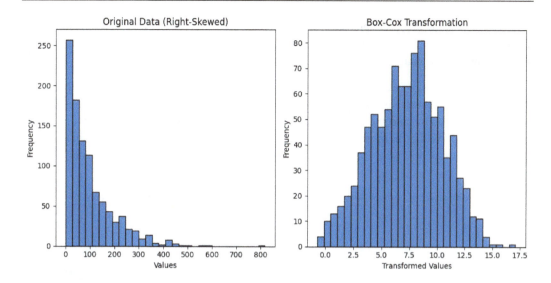

Figure 9.4: Distribution before and after a Box-Cox transformation

The histogram on the left represents the original right-skewed data, while the histogram on the right displays the transformed data after applying the Box-Cox transformation. The Box-Cox transformation helps to stabilize the variance and achieve a more symmetric distribution in the transformed data. This plot demonstrates the effectiveness of the Box-Cox transformation in addressing skewness and making data more suitable for certain types of analyses and modeling tasks.

Exponential transformations

An **exponential transformation** is a data transformation technique that takes the exponential function of each data point in a dataset. Unlike the log transformation, this transformation is often used to mitigate the effects of left-skewed or negatively skewed data, where extreme values are more frequent and the tail of the distribution is longer. By applying the exponential transformation, we stretch the values toward higher magnitudes, leading to a more symmetrical distribution.

Applying an exponential distribution to a variable as a pre-modeling exercise can be beneficial in scenarios where data or a specific variable exhibits certain characteristics, related to time-to-event or waiting time phenomena. It is also particularly useful when dealing with data that follows exponential growth or decay patterns and when the underlying process has a constant hazard rate, meaning that the probability of an event occurring in the next instant is independent of the time that has passed since the last event.

Data scientists must be aware when attempting to use this technique. First, the input data should be positive. When performing exponential transformation, many use a **base of e** or the natural exponential function, and negative values when exponentiated, which can lead to complex numbers that might not be desired in many practical applications. Additionally, they need to be aware of the magnitude of their input values. Raising a number to 1,000 can lead to an extremely large result, which could cause overflow in some computational environments.

To observe the shape of an exponential variable, the histogram in *Figure 9.5* represents the data distribution of time to purchase. The *x*-axis represents the time (in days) it takes for customers to make their first purchase, and the *y*-axis represents the frequency of customers falling within each time interval. Observe that, in most cases, customers take quite some time to make a purchase. This is expected, as most people who visit a website or sign up for a newsletter don't immediately make a purchase.

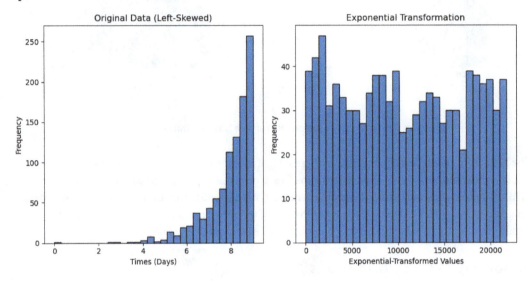

Figure 9.5: Exponential distribution of time to purchase

> **Note**
>
> While data transformations have valuable applications, it is essential to be cautious and ensure that data truly follows the characteristics of the suspected distribution before applying it. In practice, real-world data may not perfectly follow theoretical distributions, so model fit assessments and hypothesis testing are required to validate the choice of distribution.

There are more data transformations that we did not discuss, with their own unique applications, including square transformations, root transformations, Weibull transformations, and hill function transformations, but these are much rarer in generalized data science roles. We encourage you to explore these in your own time if they strike your interest!

Engineering categorical data and other features

This section will explore the handling of categorical variables in feature engineering for data science and machine learning projects. **Categorical variables** contain discrete values that represent different groups or categories. Effectively preprocessing and engineering these variables is essential to extract valuable insights and enhance the predictive power of machine learning models. We will dive into various techniques and best practices to transform categorical variables into meaningful numerical representations.

One-hot encoding

One-hot encoding is a popular technique for converting categorical variables into binary vectors. Each category is represented as a binary feature, with a value of 1 if the data point belongs to that category and 0 otherwise. For example, consider a categorical feature, `Color`, with the categories `Red`, `Blue`, and `Green`. After one-hot encoding, this feature will be split into three binary features – `Color_Red`, `Color_Blue`, and `Color_Green`. This allows machine learning algorithms to process categorical data effectively.

One-hot encoding is necessary because many machine learning algorithms cannot directly process categorical data in its original form. These values do not have a numerical relationship that algorithms can understand. Let's take another example – this time, a dataset with a categorical feature, `Gender`, with three categories – `Male`, `Female`, and `Non-Binary`. We have the following data samples:

ID	Gender
1	Male
2	Female
3	Non-Binary
4	Female
5	Male

Figure 9.6: A categorical gender dataset

After applying one-hot encoding, the `Gender` feature is transformed into binary features representing each category. For each data sample, we create new binary features – `Gender_Male`, `Gender_Female`, and `Gender_Non-Binary`. The binary features are assigned a value of 1 if the data sample belongs to that category, and 0 otherwise.

ID	Gender_Male	Gender_Female	Gender_Non-Binary
1	1	0	0
2	0	1	0
3	0	0	1

ID	Gender_Male	Gender_Female	Gender_Non-Binary
4	0	1	0
5	1	0	0

Figure 9.7: A one-hot-encoded gender dataset

Before one-hot encoding, the Gender feature is in its original categorical form, with strings representing the categories. However, machine learning algorithms require numerical data for processing. After one-hot encoding, each category is converted into its own binary feature, creating new binary columns that capture the presence or absence of each category for each data sample.

In Python, you can achieve this using the following code:

```
# Create DataFrame
df = pd.DataFrame(data)

# Perform One-Hot Encoding
df_encoded = pd.get_dummies(df, columns=['Gender'])

print("Original DataFrame:")
print(df)
print("\nOne-Hot Encoded DataFrame:")
print(df_encoded)
```

In the code, we first create a DataFrame, df, with the example data that contains a Gender column, with categorical values. Then, we use the pd.get_dummies() function from Pandas to perform one-hot encoding on the Gender column. This function automatically identifies the unique categories in the Gender column and creates new binary columns for each category.

Label encoding

Label encoding is another technique used to convert categorical data into a numerical format. Unlike one-hot encoding, which creates binary features for each category, label encoding assigns a unique numerical label to each category in the original categorical variable. The numerical labels are sequential integers, starting from 0 for the first category, 1 for the second category, and so on. For example, consider a categorical feature, Size, with the categories Small, Medium, and Large. After label encoding, the categories can be represented as 0, 1, and 2, respectively. Label encoding can be useful when there is an inherent order or ranking among the categories.

Here is another example:

```
import pandas as pd

# Example Data
```

```
data = {'ID': [1, 2, 3, 4, 5],
        'Color': ['Red', 'Blue', 'Green', 'Red', 'Green']}

# Create DataFrame
df = pd.DataFrame(data)

# Perform Label Encoding
color_mapping = {'Red': 0, 'Blue': 1, 'Green': 2}
df['Color_LabelEncoded'] = df['Color'].map(color_mapping)

print("Original DataFrame:")
print(df)
```

Here is the outcome:

ID	Color	Color_LabelEncoded
1	Red	0
2	Blue	1
3	Green	2
4	Red	0
5	Green	2

Figure 9.8: A label-encoded color dataset

In the example, we used label encoding to convert the Color categorical variable into the Color_LabelEncoded numeric feature. The Red, Blue, and Green categories are replaced with numerical labels 0, 1, and 2, respectively.

So, what's the difference between one-hot encoding and label encoding?

- **The number of features**: One-hot encoding creates binary features equal to the number of unique categories, while label encoding creates only one numerical feature.

- **Numerical representation**: One-hot encoding represents each category with a separate binary feature, where a value of 1 indicates the presence of that category. Label encoding represents each category with a unique integer label.

- **Handling high cardinality**: One-hot encoding is suitable for categorical variables with low cardinality (a few unique categories), as it creates a binary feature for each category. For high cardinality categorical variables, one-hot encoding can lead to an explosion in the number of features, making it computationally expensive. In contrast, label encoding handles high cardinality efficiently, as it uses a single integer for each category.

Target encoding

Target encoding, also known as mean encoding, is a technique that leverages the target variable's information to transform categorical features into numeric representations. Instead of replacing the categories with numerical labels, target encoding replaces each category with the average value of the target variable for that category.

For example, consider a categorical feature, `City`, with the categories `New York`, `Los Angeles`, and `Chicago`. After target encoding, each category will be replaced with the average target value for that city, such as 0.23, 0.18, and 0.32, respectively. Target encoding can be particularly useful when dealing with high-cardinality categorical variables.

Here's how target encoding works:

```
import pandas as pd

# Example Data
data = {'ID': [1, 2, 3, 4, 5],
        'City': ['Indianapolis', 'Detroit', 'Chicago', 'Detroit',
'Indianapolis'],
        'Target': [0.8, 0.6, 0.9, 0.7, 0.75]}

# Create DataFrame
df = pd.DataFrame(data)

# Perform Target Encoding
city_target_mean = df.groupby('City')['Target'].mean()
df['City_TargetEncoded'] = df['City'].map(city_target_mean)

print("Original DataFrame:")
print(df)
```

Here are the results:

ID	City	Target	City_TargetEncoded
1	Indianapolis	0.80	0.775
2	Detroit	0.60	0.650
3	Chicago	0.90	0.900
4	Detroit	0.70	0.650
5	Indianapolis	0.75	0.775

Figure 9.9: A target-encoded city dataset

In the preceding example, we used target encoding to convert the `City` categorical variable into the `City_TargetEncoded` numeric feature. The target encoding process calculated the average target value for each category (`Indianapolis`, `Detroit`, and `Chicago`) and replaced each category with its corresponding mean target value.

At this point, you may be wondering, when should you use one-hot encoding versus label encoding versus target encoding? There's not always a clear answer, but here are some things to consider:

- **One-hot encoding**: Use one-hot encoding when dealing with categorical variables with low cardinality and no inherent ordinality. One-hot encoding is essential when you want to avoid introducing any ordinal relationships or implied numerical order between categories. It is also useful when dealing with machine learning algorithms that do not handle categorical data directly.

- **Label encoding**: Use label encoding when dealing with categorical variables with inherent ordinality. In cases where the categories have a natural order or ranking, label encoding can capture this information effectively. Label encoding is also efficient when handling high cardinality categorical variables, as it reduces the number of features compared to one-hot encoding.

- **Target encoding**: Consider target encoding when dealing with categorical variables that show a strong relationship with the target variable. Target encoding can capture the average target value for each category, making it useful for generating informative numerical representations of categorical data. However, it is essential to be cautious with target encoding to avoid overfitting and data leakage. Target encoding can be particularly useful for high cardinality categorical variables, as it can provide meaningful numerical representations without introducing a large number of new features.

Calculated fields

In this section, we will explore the concept of creating calculated fields as a powerful technique for feature engineering in data science and machine learning projects. **Calculated fields** involve generating new features by applying mathematical operations, combining existing features, or extracting meaningful information from raw data. The process of crafting calculated fields empowers data scientists to capture intricate patterns, relationships, and domain-specific insights that might not be evident in the original dataset.

Well-crafted calculated fields are often superior to existing raw features. New features, when carefully designed, can capture complex patterns and relationships, making the machine learning model more robust and accurate. We will address the risk of overfitting and data leakage and provide guidelines for feature selection and evaluation. For example, a calculated field such as `Days since Last Purchase` may be more insightful than `Purchase Date` in predicting customer behavior.

Calculated fields enable us to extract complex relationships and hidden patterns, making machine learning models more effective in capturing the underlying structure of data.

There are the various types of calculated fields:

- **Mathematical operations**: These include addition, subtraction, multiplication, division, and exponentiation to create new features. These operations can help derive ratios, percentages, or other meaningful indicators that may reveal important trends in data.

  An example is calculating price per quantity for an e-commerce dataset by dividing the `Price` feature by the `Quantity` feature.

- **Aggregating and grouping**: Aggregation involves summarizing data by grouping it based on certain categorical features and computing statistics, such as mean, sum, and median, for each group. It can lead to insightful new features by capturing the collective behavior of data within specific groups.

  An example is computing the average revenue per customer by grouping customers based on their customer ID, and aggregating the `Revenue` feature.

- **Time-based calculations**: These are commonly used in time-series data or scenarios that include temporal patterns, time lags, rolling averages, and other time-related transformations that can capture trends and seasonality in data.

  An example is creating a 7-day rolling average feature for a sales dataset to identify trends and smooth out short-term fluctuations.

- **Interaction terms and polynomial features**: Interaction terms and polynomial features are important for capturing non-linear relationships between features. Combining features can reveal interactions that significantly impact the target variable.

  An example is adding an interaction term by multiplying the `Age` feature by the `Income` feature to capture the combined effect on the target variable, such as purchasing power.

- **Text and NLP-based calculations**: For datasets containing text data, feature engineering techniques using **Natural Language Processing** (**NLP**) can come in handy. This includes text vectorization, text extraction, concatenation, word counts, and a host of other NLP transformations to derive meaningful features from textual information.

  An example is extracting the sentiment score from customer reviews and using it as a feature in a sentiment analysis model.

- **Domain-specific calculations**: Domain-specific feature engineering is extremely common and what makes data scientists stand out. Expert knowledge plays a crucial role in generating relevant calculated fields.

  An example is, in the healthcare domain, calculating a `BMI (Body Mass Index)` feature based on a person's weight and height.

Assessment

Why is one-hot encoding preferred over label encoding for categorical variables with no inherent order?

Answer

One-hot encoding is preferred over label encoding when dealing with categorical variables without any inherent order because label encoding may introduce unintended ordinal relationships between categories. For example, if we encode the colors Red, Blue, and Green as 0, 1, 2, respectively, using label encoding, it may suggest that Green is somehow greater or more significant than Red, which is not the case. One-hot encoding, conversely, avoids this issue by creating separate binary features for each category, ensuring that there are no implied relationships between them.

Assessment

What are some potential challenges associated with using label encoding for categorical variables with no ordinality?

Answer

When using label encoding for categorical variables with no inherent order, some machine learning algorithms may interpret the numerical labels as continuous values and assume a natural ordering between the categories. This could lead to incorrect results, as the numerical labels are purely nominal and do not carry any meaningful numerical relationships. Additionally, if the range of label values is large, algorithms may give higher importance to categories with higher label values, even if such relationships do not exist in reality.

Assessment

How can target encoding help improve predictive accuracy in machine learning models?

Answer

Target encoding can help improve predictive accuracy by encoding categorical variables with the average target value for each category. This can capture category-specific information, especially in scenarios where the target variable exhibits distinct behavior across different categories. By incorporating this information as a numeric representation, machine learning models can learn to differentiate between the categories more effectively and make more informed predictions. However, it is important to be cautious with target encoding to avoid overfitting and data leakage, as target encoding may lead to overestimation of performance if not handled properly.

Performing feature selection

Feature selection is a critical step in the machine learning pipeline aimed at identifying the most relevant and informative features from the original dataset. By carefully selecting features, data scientists can improve model performance, reduce overfitting, enhance model interpretability, and decrease computational complexity.

Feature selection helps to focus a model on the most impactful features, making it more interpretable and reducing the risk of overfitting. In this section, we will explore scenarios where using all available features can lead to the "curse of dimensionality" and why selecting relevant features is crucial to mitigate this issue.

Types of feature selection

There are three main categories of feature selection techniques:

- **Filter methods**: These methods rank features based on statistical metrics such as correlation, mutual information, or variance. They are computationally efficient and independent of the chosen machine learning model.

- **Wrapper methods**: Wrapper methods assess feature subsets using a specific machine learning model's performance as an evaluation metric. They are computationally expensive but can lead to optimal feature subsets for specific models.

- **Embedded methods**: Embedded methods incorporate feature selection into the model training process. These methods assess feature importance during model training and eliminate less relevant features automatically.

When it comes to feature selection, one challenge for both the filter and wrapper methods is that a data scientist using these methods will need to set a threshold on the end number of features they want, or limits on how large of a performance change they are willing to accept. There isn't a universal answer to the question of how to set a threshold, and it is often situational-based. However, thinking about how much it might cost to gather and store your data could be a helpful guide. In general, you want to use the least number of features to obtain the same amount of performance from your machine learning model. Therefore, let's look at our first feature selection method, which uses the wrapper technique to select features.

Recursive feature elimination

Recursive Feature Elimination (RFE) is a wrapper method for feature selection that works iteratively to identify the most important features in a dataset. It starts by training a machine learning model on the entire feature set and ranks the features based on their importance scores. The least important feature(s) are then removed, and the model is retrained on the reduced feature set. This process is repeated until the desired number of features is reached.

RFE is particularly useful when a machine learning model provides feature importance rankings, such as decision trees or linear regression. By eliminating less important features at each iteration, RFE aims to find an optimal subset of features that maximizes the model's performance.

Without getting too ahead of ourselves, here is how you implement RFE in Python using the RFE package:

```
from sklearn.datasets import load_iris
from sklearn.feature_selection import RFE
from sklearn.linear_model import LogisticRegression

# Load the Iris dataset
data = load_iris()
X = data.data
y = data.target

# Create a logistic regression model
model = LogisticRegression()

# Initialize RFE and specify the number of features to select
rfe = RFE(model, n_features_to_select=2)

# Fit RFE on the data
rfe.fit(X, y)

# Get the selected features
selected_features = rfe.support_
print('Selected Features:', selected_features)

# Get the feature ranking
feature_ranking = rfe.ranking_
print('Feature Ranking:', feature_ranking)
```

In the code block, after declaring our imports, we first load in the *Iris* dataset, which focuses on classifying flowers based on different physical characteristics. Once we load the dataset, we create an instance of a logistic regression classifier, which will take the input data and attempt to learn how to classify the different flowers.

The key in this code block is that we use an instance of the RFE function to wrap our logistic model in. We state that we want to select the top two features in the dataset. This will take care of the RFE feature selection process for us. Finally, we are left with the two most important features.

Now that we have looked at a feature selection process that uses the wrapper technique, let's look at one that uses the embedded technique.

L1 regularization

L1 regularization, also known as the **Least Absolute Shrinkage and Selection Operator (LASSO)**, is an embedded feature selection method that combines feature selection and regularization during model training. In the LASSO, the linear regression coefficients are penalized based on the absolute values of the coefficients. This penalty encourages some of the coefficients to be exactly zero, effectively performing feature selection by excluding irrelevant features.

The LASSO is particularly effective when dealing with high-dimensional datasets where the number of features is much larger than the number of samples. By driving some feature coefficients to zero, the LASSO automatically selects the most relevant features and performs a form of dimensionality reduction. It helps to enhance model interpretability and generalization while avoiding the risk of overfitting.

Tree-based feature selection

Tree-based models, such as random forest and gradient boosting, can provide valuable feature importance scores. These scores indicate the relative importance of each feature in predicting the target variable. Tree-based feature selection involves using these importance scores to rank features and selecting the top-performing ones. We will talk more about these models in the chapter on machine learning.

Tree-based feature selection is computationally efficient and applicable to both classification and regression tasks. It is especially useful for identifying relevant features in datasets with a mix of categorical and numerical variables. Additionally, tree-based models can handle nonlinear relationships, making them suitable for datasets with complex interactions between features. We will discuss tree-based models in *Chapter 10*.

The variance inflation factor

Collinear features (or **multicollinear features** for 3 or more variables) refer to variables that are highly correlated with each other. Such features can introduce redundancy in the dataset and impact model interpretability. Additionally, collinearity can lead to unstable model coefficients, making it challenging to identify the true impact of individual features on the target variable.

Techniques such as the **Variance Inflation Factor** (**VIF**) can be used to detect collinearity between features. A high VIF score for a feature indicates strong multicollinearity, while a VIF close to 1 indicates no collinearity. To address collinearity, data scientists may choose to remove one of the highly correlated features or perform dimensionality reduction, using techniques such as PCA to create uncorrelated principal components.

Handling collinear features is crucial for maintaining model stability and ensuring that feature selection and feature importance rankings are based on independent and informative features, leading to more accurate and interpretable models.

Here is how to implement a VIF in Python:

```python
import pandas as pd
import numpy as np
from sklearn.datasets import fetch_california_housing
from statsmodels.stats.outliers_influence import variance_inflation_
factor

# Load the California Housing dataset
data = fetch_california_housing()
X = pd.DataFrame(data.data, columns=data.feature_names)
y = data.target

# Function to calculate VIF for each feature
def calculate_vif(X):
    vif = pd.DataFrame()
    vif['Feature'] = X.columns
    vif['VIF'] = [variance_inflation_factor(X.values, i) for i in
range(X.shape[1])]
    return vif

# Calculate VIF for the entire feature set
vif = calculate_vif(X)
print(vif)
```

In the code block, we first load the *California Housing* dataset. In this dataset, we attempt to predict the median house value for California districts, based on information such as the average number of rooms per household and the median house age. After loading the dataset, we create a function to compute the VIF for each feature. After running the function, we print out the results. From this point, we can create a filter to remove those features, with a VIF greater than some threshold we set based on our project.

Working with imbalanced data

In this section, we will explore the challenges posed by imbalanced datasets in machine learning and various methods to effectively address this issue. **Imbalanced data** refers to datasets where one class (the minority class) is significantly underrepresented compared to another class (the majority class). The class imbalance can lead to biased and suboptimal model performance, as models tend to favor the majority class, making accurate predictions for the minority class challenging. We will delve into the consequences of imbalanced data and several techniques to handle imbalanced datasets for improved model performance.

Understanding imbalanced data

Since models prioritize the majority class, there are serious consequences of imbalanced data on model training and evaluation.

In the context of imbalanced datasets in machine learning, the **majority class** refers to the class that has a significantly larger number of instances or observations compared to the other class(es) in the dataset. It is the class that dominates the dataset in terms of its representation, and as a result, machine learning models trained on imbalanced datasets may be biased toward predicting this majority class more frequently.

Conversely, the **minority class** refers to the class that has a relatively smaller number of instances or observations compared to the other class(es) in the dataset. This class is underrepresented and may have fewer data points available for the model to learn from. As a consequence, machine learning models may struggle to correctly predict this minority class and may have lower accuracy, recall, and precision for this class.

For example, consider a binary classification problem where we try to predict whether an email is spam. If the dataset contains 900 non-spam (not spam) emails and only 100 spam emails, the non-spam class is the majority class and the spam class is the minority class. In this scenario, the dataset is imbalanced, due to the significant difference in the number of instances between the two classes.

As you may have guessed, we can't avoid these scenarios entirely because many business problems are based on imbalanced datasets. Consider e-commerce, where you model website clicks when the site receives thousands of visits daily. In most cases, a click is very rare. Without adjustments for the imbalanced class, the model will likely prioritize the majority class, leading to high accuracy but poor recall and precision for the minority class.

Treating imbalanced data

Imbalanced data can have several consequences on machine learning models. This subject could take an entire book to explain, but here are some approaches for further investigation. Your takeaway here should be the depth and general logic behind imbalanced dataset remedies. Showing your knowledge in this area will show your level of understanding in pre-modeling practices and considerations.

Some remedies for imbalanced data include the following:

- **Using different evaluation metrics**: Use performance metrics that are more appropriate for imbalanced datasets than simple accuracy. Metrics such as precision, recall, F1-score, and **area under the receiver operating characteristic curve** (**AUC-ROC**) are better suited for evaluating model performance on imbalanced data.

- **Over-sampling**: This involves generating synthetic samples for the minority class to increase its representation.

- **Under-sampling**: This involves randomly removing samples from the majority class to decrease its dominance.

- **Random under-sampling and over-sampling**: Python offers libraries such as `imbalanced-learn` to implement this technique.

- **Synthetic minority over-sampling technique (SMOTE)**: SMOTE is a popular over-sampling technique that generates synthetic samples by interpolating between neighboring samples of the minority class. You can use `SMOTE` from the `imblearn.over_sampling` package in Python.

- **Ensemble methods**: Ensemble methods, such as Random Forest and Gradient Boosting, can handle imbalanced data effectively due to their inherent robustness.

- **Cost-sensitive learning**: Cost-sensitive learning is an approach that assigns different misclassification costs to different classes, guiding a model to prioritize the minority class.

- **Using anomaly detection**: Anomaly detection techniques can be useful in handling imbalanced data by identifying and classifying rare instances as anomalies. These algorithms include Isolation Forest and One-Class SVM.

Reducing the dimensionality

In this section, we will explore the concept of **dimensionality reduction**, a critical technique in machine learning and data analysis that aims to reduce the number of features or variables in a dataset while preserving essential information. High-dimensional datasets often suffer from the "curse of dimensionality," leading to increased computational complexity and potential overfitting. Dimensionality reduction methods help to transform data into a lower-dimensional space, enabling easier visualization, improved model performance, and enhanced interpretability.

Here, we will delve into various dimensionality reduction techniques, and their applications, and provide code examples in Python to implement them effectively.

Principal component analysis

Principal Component Analysis (PCA) is a widely used linear dimensionality reduction technique that projects data onto orthogonal axes to capture the maximum variance in the first principal components.

PCA is a popular linear dimensionality reduction technique used to transform high-dimensional data into a lower-dimensional space. It achieves this by identifying the principal components, which are orthogonal directions that capture the maximum variance in the data. The first principal component represents the direction of the highest variance, the second principal component represents the second highest variance, and so on. By selecting a reduced number of principal components, we can project the data onto a lower-dimensional subspace while retaining the most relevant information.

PCA is widely used for data visualization, feature extraction, and noise reduction. It helps in identifying the main patterns and trends in data, simplifying data representation, and speeding up machine learning algorithms by reducing computational complexity. However, PCA assumes linearity in the data and may not perform well on complex non-linear relationships.

To implement PCA in Python, we can use libraries such as NumPy and `scikit-learn`. Here's a step-by-step guide:

```
import numpy as np
from sklearn.decomposition import PCA

# Sample data
X = np.random.rand(100, 5)

# Create a PCA instance and fit the data
pca = PCA(n_components=2)
X_pca = pca.fit_transform(X)

# Print the explained variance ratio of the two principal components
print("Explained Variance Ratio: ", pca.explained_variance_ratio_)
```

The code starts by generating a dataset that has five features or dimensions. Then, we use the PCA model to select the top two components. Finally, we look at how much variance is explained by the top two PCA components. So, we have reduced the number of dimensions from five down to two. Data scientists look to find the how many components are needed to explain most of the variance in the dataset.

Singular value decomposition

Single Value Decomposition (SVD) is a fundamental matrix factorization technique that plays a key role in PCA. SVD is used to decompose a matrix into three matrices, U, Σ, and V^T. U and V^T are orthogonal matrices, while Σ is a diagonal matrix containing the singular values.

In PCA, SVD is applied to the centered data to obtain the principal components and explained variance. In Python, we can run SVD like this:

```
import numpy as np

# Sample data
X = np.random.rand(100, 5)

# Center the data
X_centered = X - X.mean(axis=0)

# Perform SVD
```

```
U, S, Vt = np.linalg.svd(X_centered, full_matrices=False)

# Reduce dimensionality to two dimensions using the first two
X_svd = np.matmul(U[:, :2], np.diag(S[:2]))
```

Here, we reduce our initial data from five columns down to two columns.

t-SNE

t-distributed Stochastic Neighbor Embedding (**t-SNE**) is a non-linear dimensionality reduction technique used when data has complex non-linear relationships. t-SNE aims to preserve local and global structures in the data.

You can apply t-SNE in Python using libraries such as `scikit-learn`:

```
import numpy as np
from sklearn.manifold import TSNE

# Sample data
X = np.random.rand(100, 5)

# Create a t-SNE instance and fit_transform the data
tsne = TSNE(n_components=2)
X_tsne = tsne.fit_transform(X)
```

This code takes our randomly generated data and reduces it to two dimensions. It is important to know as a data scientist that this algorithm needs to compute pair-wise distances for each data point in the dataset. Therefore, the time it takes to run increases with the size of the dataset. So, this may not be a great candidate when you have a large amount of data.

Autoencoders

Autoencoders are neural networks used for unsupervised representation learning and non-linear dimensionality reduction. They consist of an encoder that compresses data into a lower-dimensional representation and a decoder that reconstructs the data from the compressed representation.

Here is an example of how to use an autoencoder in Python for dimensionality reduction:

```
import numpy as np
from keras.layers import Input, Dense
from keras.models import Model

# Sample data
X = np.random.rand(100, 5)
```

```
# Define the autoencoder architecture
input_layer = Input(shape=(5,))
encoded = Dense(2, activation='relu')(input_layer)
decoded = Dense(5, activation='sigmoid')(encoded)

autoencoder = Model(input_layer, decoded)

# Compile the autoencoder
autoencoder.compile(optimizer='adam', loss='mean_squared_error')

# Train the autoencoder
autoencoder.fit(X, X, epochs=50, batch_size=32)

# Obtain the lower-dimensional representation (encoder part of the
autoencoder)
encoder = Model(input_layer, encoded)
X_autoencoder = encoder.predict(X)
```

In the previous code block, we created a model using Keras, where we designed it to have two layers. The first layer encodes the input data into two dimensions. The last layer decodes that information back. The model attempts to learn a representation, where it first encodes the data and decodes it perfectly. Once the training is done, we can throw away the last layer and only use the encoder portion for dimensionality reduction. This is just another tool in the data scientist's tool bag when they need to perform dimensionality reduction.

Summary

In this comprehensive chapter, we covered essential concepts in pre-modeling data for analytics and feature engineering. Mastering these techniques is vital for data scientists to effectively handle real-world datasets and build accurate machine learning models.

Understanding techniques such as data min-max scaling, z-score scaling, and feature engineering can enhance model performance; transformations such as logarithmic, Box-Cox, and exponential help reshape data for better algorithm compatibility; dimensionality reduction methods such as PCA and t-SNE simplify and visualize data and aid in effective model building; and handling imbalanced data with resampling and ensemble techniques ensure balanced datasets and unbiased predictions.

Additionally, we covered feature engineering techniques, including one-hot encoding, label encoding, and target encoding. These techniques allow us to craft new and informative representations of data. Feature engineering involves selecting, transforming, and creating features that best capture the underlying patterns and relationships within data so that we ensure robust and accurate models.

In the next chapter, we will focus on machine learning algorithms.

10
Mastering Machine Learning Concepts

It's time to give yourself a very generous pat on the back because you've officially arrived at the chapter on machine learning concepts. Take a moment to appreciate how far you've come, as well as all the preliminary information in the earlier chapters it takes to truly grasp this chapter. Many learners do themselves a disservice by jumping right into machine learning without first understanding its underlying principles (for example, statistics) and preliminary tasks (for example, data wrangling or pre-modeling), so this puts you ahead of the curve as someone well-equipped to understand the inner workings of machine learning algorithms and how and when to use them.

Throughout this chapter, we will cover a wide array of machine learning topics, providing you with the foundation needed to understand the intricacies of various algorithms and techniques. Our journey will begin with a detailed examination of the machine learning workflow – a step-by-step process that data scientists follow when tackling real-world problems. Then, with the groundwork laid, we will explore the world of supervised learning, one of the fundamental branches of machine learning. After that, we will transition to unsupervised learning, where we will explore the world of clustering algorithms. Furthermore, we will discuss various evaluation metrics to gauge model effectiveness and explore the bias-variance trade-off – a fundamental concept that highlights the delicate balance between model complexity and generalization. Finally, we will explore cross-validation and hyperparameter tuning methods to ensure our models perform optimally.

After completing this chapter, you will be able to critically analyze the strengths and weaknesses of different machine learning models, allowing you to make informed decisions when selecting the most appropriate algorithm for specific tasks. With hands-on coding examples and real-world use cases throughout this chapter, you will gain practical experience and confidence in applying machine learning concepts to tackle data-driven challenges.

So, in this chapter, we will cover the following topics:

- Introducing the machine learning workflow
- Getting started with supervised learning
- Getting started with unsupervised learning
- Summarizing other notable models
- Understanding the bias-variance trade-off
- Tuning with hyperparameters

Introducing the machine learning workflow

If you're a data scientist preparing for a technical interview, understanding the machine learning workflow is non-negotiable. **Machine learning** is concerned with the design and application of algorithms and techniques that allow computers to learn patterns that are often applied to solve business problems.

At its core, the workflow consists of several key stages, beginning with a well-defined problem statement and culminating in the application of a model trained on unseen data. Each stage, whether it's selecting the appropriate model, tuning hyperparameters, or making predictions, serves as an essential step in the data science process. Mastery of these stages not only sharpens your technical acumen but also equips you with the systematic thinking required to tackle a wide range of data-related problems:

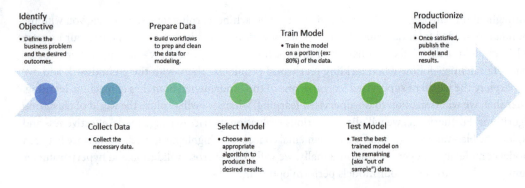

Figure 10.1: Workflow for machine learning projects

The importance of the machine learning workflow extends beyond just the theoretical understanding of algorithms. In interviews and practical settings alike, you'll often be evaluated on your ability to articulate the rationale behind each choice you make – why you chose a specific model, how you tuned it, and how you assessed its performance.

We will delve deep into these areas, covering common models, their strengths and weaknesses, and fine-tuning techniques. You'll also learn about model evaluation metrics and how to interpret them, ensuring that you're not just following steps, but also understanding the implications of each decision you make. By the end of this section, you'll be better prepared to articulate and execute a robust machine learning workflow, setting you apart in any data science interview.

Problem statement

At the heart of every data science endeavor lies a well-defined **problem statement**. This initial step involves understanding the problem at hand, identifying goals, and outlining the data needed for analysis. Clear problem formulation helps set the direction for subsequent stages, ensuring a focused and purpose-driven approach.

Model selection

Selecting an appropriate machine learning model is a critical decision in the data science workflow. Depending on the nature of the problem – whether it involves classification, regression, clustering, or other tasks – careful consideration is given to the strengths and weaknesses of various models. The **model selection** stage requires a deep understanding of algorithms and their applicability to the problem context.

Model tuning

Once a model has been chosen, the **model-tuning** process comes into play. This stage involves optimizing hyperparameters to achieve the best possible model performance. Techniques such as grid search, random search, and Bayesian optimization are employed to fine-tune the model. We will review each of these techniques in more detail later in this chapter, but for now, just note that they are different methods for trying different combinations of model hyperparameters to find the set that gives the best overall model performance. Model tuning balances overfitting and underfitting, ensuring the model generalizes well to unseen data.

Model predictions

The culmination of the data science workflow is applying the trained model to new, unseen data. This **prediction** phase involves leveraging the model's learned patterns and relationships to make accurate predictions or classifications. It's the moment where the fruits of the entire data science process come to fruition as the model's effectiveness is put to the test on real-world data.

Of course, there are more stages than the ones mentioned here, including communicating with stakeholders, tracking experiments, and monitoring data drift, but for the sake of this chapter, these are the primary areas on which we will focus. Particularly, we will review common models in data science, including how they work, their assumptions, common pitfalls, implementation examples, model evaluation, and tuning.

Getting started with supervised machine learning

Supervised learning is a type of machine learning where the algorithm learns from a labeled dataset, which consists of input features and their corresponding target variables or labels. These labels are the "response variable," "target variable," or "output variable" – in other words, the thing you are trying to predict.

There are two types of supervised modeling that we will focus on:

- Regression
- Classification

Let's take a closer look at them.

Regression versus classification

Regression is a specific type of supervised learning where the goal is to predict continuous numerical values. In a regression task, the algorithm learns a mapping between input features and a continuous target variable. The output of the regression model is a continuous value, which can represent quantities such as price, temperature, sales, or any other real-valued quantity. Linear regression and polynomial regression are common examples of regression algorithms that are used to model relationships between variables in a continuous setting.

For example, imagine that you are performing an analysis of how prepared a person is for retirement. In this case, you are looking to predict how much a person has saved for retirement based on demographic data. *Figure 10.2* shows an example record for input features that are mapped to a continuous output variable, the retirement balance:

Age	Gender	Wage	Years in Field	Debt	Retirement Balance
40	1	54,000	20	10,000	64,000

Figure 10.2: Regression data example

Here, each row represents a subject's description and their current retirement fund balance. This is an example of regression problem data because the output variable (the thing we are trying to predict) is a continuous target variable. Because we know that the target variables in regression problems are continuous, we also know how to evaluate the model's performance.

Here are some common regression model evaluation metrics:

- **Mean squared error** (**MSE**): This metric takes the average of the squared differences between the predicted and actual values and penalizes large errors heavily. If the dataset contains outliers, they will have a disproportionate impact.

- **Root mean squared error** (**RMSE**): This metric takes the square root of the mean squared error. It is also sensitive to outliers.

- **Mean absolute error** (**MAE**): This metric is the average of the absolute differences between the predicted and actual values.

Now, let's compare regression, which only outputs continuous variables, to classification.

Classification is another version of supervised learning that focuses on predicting categorical labels or classes for a given set of input features. In a classification task, the algorithm learns to differentiate between different categories based on patterns in the training data. The output of a classification model is a discrete label representing the predicted class to which the input data belongs. Common examples of classification problems include email spam detection (binary classification), handwritten digit recognition (multiclass classification), and sentiment analysis (multiclass classification).

Consider *Figure 10.3*, which references back to our retirement analysis. However, this time, it shows an example record for input features that are mapped to a categorical output variable – whether the person is ready for retirement. In this scenario, we will assume that 1 equals yes and 0 equals no:

Age	Gender	Wage	Years in Field	Debt	Ready for Retirement?
40	1	54,000	20	10,000	1

Figure 10.3: Classification data example

Similar to regression, if you know that your problem involves predicting a categorical variable, you have the following model performance evaluation metrics available to you:

- **Accuracy**: This metric measures the percentage of predictions that are correct.

- **Precision**: Precision is the percentage of predicted positive classes that are positive. This can be an important metric if your dataset is imbalanced.

- **Recall** (that is, the **sensitivity** or **true positive rate**): This metric is the percentage of actual positives that are correctly predicted as positive and is complementary to precision.

- **Specificity** (that is, the **true negative rate**): Specificity is the percentage of actual negatives that are correctly predicted as negative. This is important in cases where false positives are costly, such as in medical diagnosis.

- **F1 score**: This metric combines both precision and recall into one metric, and is a good compromise between both.

- **Area under the receiver operating characteristic** (**AUC**): This metric is a measure of how well the model can distinguish between positive and negative classes. The AUC is not affected by class imbalance.

We've now reviewed two types of supervised learning in regression and classification. When working through your technical interview, the interviewer will expect you to know if you are using a classification or regression model. The target variable is the key to this decision point. Thus, if you can identify the format of the target variable, you can identify the types of models that will best fit your data science problem, as well as the best evaluation metrics to use.

In the rest of this section, we will provide examples of both regression and classification models, their assumptions, and their pros and cons.

Linear regression – regression

Linear regression is a fundamental and widely used statistical method in the field of data analysis and machine learning, providing a simple yet powerful framework for modeling the relationship between one or more independent variables and a dependent variable. The goal of linear regression is to find the best-fitting linear relationship that describes the data, enabling us to make predictions and gain insights into the underlying patterns.

How it works

Linear regression works by fitting a linear equation to the observed data. The linear equation has the following form:

$$Y_i = \beta_0 + \beta_1 X_i$$

Here, Y is the dependent variable that we want to predict, $\beta 0$ is our intercept or constant, and $\beta 1$ is our slope.

The goal of linear regression is to estimate the coefficients; this involves finding the line (or hyperplane in higher dimensions) that best fits the data points. The estimation of coefficients is typically done using optimization techniques such as the **ordinal least squares** method, which aims to find the coefficients that minimize the sum of squared residuals.

Assumptions

Before delving into the intricacies of linear regression, it's important to understand the assumptions that underlie its usage. These assumptions ensure the validity and reliability of the results obtained from linear regression models. The main assumptions are as follows:

- **Linearity**: The relationship between the independent variables and the dependent variable is assumed to be linear. This means that a change in the independent variables leads to a proportional change in the dependent variable.

- **Independence**: The observations or data points are assumed to be independent of each other. In other words, the value of the dependent variable for one observation does not depend on the values of the dependent variable for other observations.

- **Homoscedasticity**: The variance of the errors (the differences between observed and predicted values) is constant across all levels of the independent variables. This assumption ensures that the model's predictions are equally accurate across the entire range of the data. When a dataset exhibits **homoscedasticity**, it implies that the variance of the errors is the same for all values of the predictor variable(s). The opposite of homoscedasticity is **heteroscedasticity**, which occurs when the variability of the errors or data points changes systematically as you move along the range of the independent variable(s).

- **Normality of residuals**: The residuals (the differences between observed and predicted values) should follow a normal distribution. Deviations from normality can impact the accuracy of statistical tests and confidence intervals.

Common pitfalls

While linear regression is a valuable tool, there are several pitfalls to watch out for:

- **Violating assumptions**: Failing to meet the assumptions of linear regression can lead to inaccurate results and misleading interpretations.

- **Outliers**: Outliers can disproportionately influence the model's coefficients, leading to an erroneous fit.

- **Multicollinearity**: **Multicollinearity** is when independent variables are highly correlated. When this happens, it can be difficult to discern their individual effects on the dependent variable.

- **Overfitting**: Adding too many variables or polynomial terms can result in overfitting, where the model captures noise in the data rather than the underlying pattern.

Regularization regression

Regularized regression (specifically **L1** (**lasso**) and **L2** (**ridge**) regression) is an extension of linear regression that addresses some of its limitations. While linear regression aims to find the best-fitting line or hyperplane, regularized regression introduces penalty terms to prevent overfitting and improve model performance. L1 and L2 regularization methods add complexity control by imposing constraints on the coefficients of the regression equation.

Regularization adds a penalty term to the linear regression objective function, which discourages the model from assigning excessively large coefficients to the features. L1 regularization adds the absolute values of the coefficients as penalties, leading to some coefficients being exactly zero, while L2 regularization adds the squared values of the coefficients, which enforces smaller but non-zero coefficients:

L1 regularization (lasso):

$$minimize\left(\sum\nolimits_{i=1}^{N}\left(y_i - \beta_0 - \sum\nolimits_{j=1}^{P}\beta_j x_{ij} \right)^2 + \lambda \sum\nolimits_{j=1}^{P}|\beta_j| \right)$$

L2 regularization (ridge):

$$minimize\left(\sum\nolimits_{i=1}^{N}\left(y_i - \beta_0 - \sum\nolimits_{j=1}^{P}\beta_j x_{ij} \right)^2 + \lambda \sum\nolimits_{j=1}^{P}\beta_j^2 \right)$$

L1 and L2 regularization are particularly useful in the following ways:

- **Feature selection**: L1 regularization can force some coefficients to be exactly zero, effectively performing feature selection and identifying the most important variables
- **Multicollinearity management**: L2 regularization helps mitigate the effects of multicollinearity by shrinking the coefficients towards zero
- **High-dimensional data**: Regularized regression is valuable when dealing with datasets with a large number of features, preventing overfitting and improving generalization

Implementation example

Here's a simple example of how to implement linear regression using Python and scikit-learn:

```
# Import necessary libraries and prepare data
from sklearn.model_selection import train_test_split
from sklearn.datasets import fetch_california_housing
from sklearn.linear_model import LinearRegression
from sklearn.metrics import mean_squared_error
housing = fetch_california_housing()
X = housing.data
```

```
y = housing.target
X_train, X_test, y_train, y_test = train_test_split(X, y, test_
size=0.2, random_state=42)

# Train the model and compute error
linear_reg = LinearRegression()
linear_reg.fit(X_train, y_train)
y_pred = linear_reg.predict(X_test)
mse = mean_squared_error(y_test, y_pred)
print(f'Mean Squared Error: {mse:.2f}')
```

Let's take a closer look at this example:

- Initially, we run common steps for any machine learning project, first importing the needed libraries from `sklearn`.

- Then, we split our dataset into a train and test set

- Next, we initialize a `LinearRegession` model and train it using the fit method with the training dataset

- Using the trained model, we make predictions using the test dataset

- Finally, we check how good our predictions are by measuring the mean square error

Assessment

What are the key assumptions of linear regression?

Answer

The key assumptions include linearity, independence of observations, homoscedasticity, and normality of residuals.

Assessment

What are some common challenges in linear regression?

Answer

Common challenges include violating assumptions, dealing with outliers, managing multicollinearity, and avoiding overfitting.

Assessment

How do you handle multicollinearity in linear regression?

Answer

Multicollinearity can be addressed by removing correlated variables, using dimensionality reduction techniques, or applying regularization methods.

Assessment

How do L1 and L2 regularization differ? What are the benefits of regularization?

Answer

L1 regularization encourages sparsity by forcing some coefficients to be exactly zero. L2 regularization shrinks coefficients toward zero but rarely eliminates them entirely. Regularization helps manage overfitting, performs feature selection, and improves model generalization, particularly in high-dimensional datasets.

Logistic regression

Logistic regression is a widely used statistical method for classification that models the relationship between one or more independent variables and a binary outcome variable. Despite its name, logistic regression is used for classification tasks rather than regression. It estimates the probability that an instance belongs to a particular class, making it an essential tool in binary and multi-class classification problems.

How it works

Logistic regression transforms the linear combination of independent variables into a probability using the logistic function (also known as the sigmoid function):

$$y = \frac{1}{1 + e^{-(a + bx_1 + cx_2)}}$$

The function returns a value between 0 and 1, which is the probability of the occurrence of an event or a class, given the input values. An example where you might use logistic regression includes predicting if a customer might churn and leave a company's service.

Assumptions

Logistic regression relies on the following assumptions:

- **Linearity of the logit**: The log odds of the probability of the outcome variable being in a certain class is a linear combination of the independent variables

- **Independence of errors**: The errors or residuals are assumed to be independent of each other

- **Non-multicollinearity**: The independent variables should not be highly correlated with each other

- **Sufficiently large sample size**: Logistic regression works best with a sufficiently large sample size to ensure stable estimates

Common pitfalls

When working with logistic regression, it's important to be aware of potential pitfalls:

- **Imbalanced classes**: Logistic regression may perform poorly when dealing with imbalanced class distributions. Some methods to address imbalanced class distribution include resampling, adjusting the weights of a class (e.g. how important an example is when training a model), and looking at different evaluation metrics like recall.

- **Non-linear relationships**: Logistic regression assumes a linear relationship between the independent variables and the log odds of the outcome. Complex non-linear relationships may not be captured effectively.

- **Overfitting**: Including too many variables or polynomial terms can lead to overfitting, especially with limited data.

- **Multicollinearity**: Highly correlated independent variables can affect the stability and interpretability of coefficient estimates.

Implementation example

Here's a simple example of how to implement logistic regression using Python and scikit-learn:

```
# Import necessary libraries and prep dataset
from sklearn.model_selection import train_test_split
from sklearn.datasets import load_iris
from sklearn.linear_model import LogisticRegression
from sklearn.metrics import accuracy_score
iris = load_iris()
X = iris.data
# Binary classification: Setosa vs. Others
y = (iris.target == 2).astype(int)
X_train, X_test, y_train, y_test = train_test_split(X, y, test_
size=0.2, random_state=42)
```

```
# Train the Logistic Regression model and compute accuracy
logreg = LogisticRegression()
logreg.fit(X_train, y_train)
y_pred = logreg.predict(X_test)
accuracy = accuracy_score(y_test, y_pred)
print(f'Accuracy: {accuracy:.2f}')
```

Let's take a closer look at this example:

- After loading the required libraries and data for the Iris flower dataset, we split the data into training and testing datasets (the Iris dataset is often used for classification machine learning problems)

- Then, we initialize a `LogisticRegression` model and train it using the fit method with the training dataset

- Using the trained model, we make predictions using the test dataset

- Finally, we check how good our predictions are by measuring the classification accuracy

Assessment

What is the logistic function, and why is it used in logistic regression?

Answer

The logistic function (sigmoid function) maps the linear combination of features to a probability between 0 and 1, enabling classification.

Assessment

How do you address imbalanced classes in logistic regression?

Answer

Techniques to address imbalanced classes include resampling methods, adjusting class weights, and using different evaluation metrics.

k-nearest neighbors (k-NN)

k-NN is a versatile and intuitive machine learning algorithm that operates based on the proximity of data points. It can be used both for classification and regression problems.

How it works

k-NN is a lazy learning algorithm, meaning it doesn't build a distinct model during training. Instead, it memorizes the training data, and when presented with a new data point for prediction, it identifies the k nearest neighbors in the training set based on a chosen distance metric (for example, Euclidean distance). The majority class among the k neighbors determines the predicted class for the new point in classification tasks. For regression tasks, the algorithm returns the average value of the target variable among the k neighbors.

Assumptions

k-NN operates under the assumption that points in the same class or category tend to be close to each other in feature space. This makes it well-suited for cases where the underlying decision boundaries are complex and not easily separable by linear methods.

Common pitfalls

Here are some common pitfalls when using k-NN:

- **Choice of k**: Selecting an appropriate value for k is crucial. A small k might lead to noisy decisions, while a large k could result in overly smooth decision boundaries.

- **Feature scaling**: k-NN is sensitive to the scale of features. Feature scaling, such as normalization or standardization, is often necessary to ensure that all features contribute equally to distance calculations.

- **Curse of dimensionality**: In high-dimensional spaces, the "nearest" neighbors might not be truly representative, leading to decreased accuracy and increased computation time.

Implementation example

Here is an example implementation of k-NN in Python using the `KNeighborsClassifier` module in sklearn:

```
# Import necessary libraries and prep dataset
from sklearn.datasets import load_iris
from sklearn.model_selection import train_test_split
from sklearn.neighbors import KNeighborsClassifier
from sklearn.metrics import accuracy_score
iris = load_iris()
X, y = iris.data, iris.target
X_train, X_test, y_train, y_test = train_test_split(X, y, test_
size=0.2, random_state=42)

# Create a KNN classifier with k=3 and compute accuracy
knn = KNeighborsClassifier(n_neighbors=3)
```

```
knn.fit(X_train, y_train)
y_pred = knn.predict(X_test)
accuracy = accuracy_score(y_test, y_pred)
print(f'Accuracy: {accuracy:.2f}')
```

Let's take a closer look at this example:

- The Iris dataset, a popular dataset for classification problems, is loaded.

- The dataset is split into training (80% of the data) and testing (20% of the data) sets to evaluate the model's performance on unseen data.

- A k-NN classifier is created with three neighbors, and it's trained using the training data

- The model's predictions on the testing set are compared with the true labels to calculate and print the accuracy of the classifier

Assessment

What's the underlying idea behind the k-NN algorithm?

Answer

k-NN predicts the class of an instance based on the classes of its k nearest neighbors in the feature space. It assumes that similar instances have similar labels.

Assessment

How do you choose the optimal value of k in k-NN?

Answer

The choice of k is a hyperparameter. You can use techniques such as cross-validation to find the optimal k value that balances bias and variance in your predictions.

Assessment

What are the pros and cons of k-NN?

Answer

Pros include simplicity and effectiveness in capturing complex decision boundaries. Cons include sensitivity to noise and the need for efficient data structures for quick searching.

Random forest

Random forest is a versatile and powerful ensemble learning technique that's used for both classification and regression tasks. It is an extension of decision tree algorithms that addresses their limitations by combining multiple trees to create a more robust and accurate predictive model. Random forest is renowned for its ability to handle complex relationships, reduce overfitting, and provide feature importance insights.

How it works

Random forest constructs an ensemble of decision trees, each trained on a different subset of the data and considering a subset of features. **Ensemble methods** are techniques in machine learning that combine the predictions of multiple individual models to create a more robust overall prediction. The idea behind ensemble methods is to exploit the diversity of different models to improve the overall accuracy, stability, and generalization of the predictive model.

Ensemble methods are particularly effective when individual models have varying strengths and weaknesses or when they can capture different aspects of the underlying data patterns. By combining these models, ensemble methods aim to mitigate the weaknesses of individual models and produce a more reliable and accurate prediction.

Random forest offers several advantages over individual decision trees:

- **Reduced overfitting**: By averaging predictions from multiple trees, random forest mitigates the risk of overfitting and provides better generalization.

- **Robustness**: Random forest is less sensitive to noisy data and outliers compared to a single decision tree

- **Non-linearity handling**: It can capture complex nonlinear relationships between features and the target variable

- **Feature importance**: Random forest quantifies the importance of each feature, aiding in feature selection and interpretation

Random forest calculates feature importance based on how much a particular feature contributes to the overall predictive performance of the ensemble. The importance of a feature is assessed by measuring the decrease in a specific metric when the values of that feature are randomly permuted while keeping the other features constant. Beyond understanding which features are the most important in a model, a data scientist might look to optimize the performance of the model.

Random forests have several hyperparameters that allow you to customize and fine-tune the behavior of the ensemble algorithm. Adjusting these hyperparameters can impact the performance, robustness, and computational efficiency of the random forest model. Here's a list of some important random forest hyperparameters, along with explanations of what they are:

- **n_estimators**: The number of decision trees in the ensemble (forest). Increasing the number of trees generally improves performance until reaching a point of diminishing returns or overfitting.

- **max_depth**: The maximum depth of each decision tree in the forest. It limits the number of splits, helping to control model complexity and reduce overfitting.

- **min_samples_split**: The minimum number of samples required to split a node further. It prevents nodes with very few samples from being split, potentially reducing noise.

There are many other hyperparameters a data scientist might want to explore optimizing. To find a list of additional hyperparameters, you can reference the sklearn documentation at `https://scikit-learn.org/stable/modules/generated/sklearn.neighbors.KNeighborsClassifier.html`.

Assumptions

Random forest is a powerful ensemble learning algorithm that combines multiple decision trees to make predictions. Unlike some other machine learning algorithms, random forest has fewer assumptions.

However, it's important to note that while individual decision trees have certain assumptions, the ensemble method helps to mitigate the impact of these assumptions:

- **Independence of observations**: Individual decision trees assume that observations are independent of each other. While this is a common assumption in many statistical and machine learning methods, random forest's ensemble approach helps reduce the impact of violations of this assumption. The random sampling and averaging of predictions across multiple trees tend to mitigate the effects of correlated or dependent observations.

- **Linearity**: Decision trees assume that relationships between features and the target variable can be modeled with piecewise constant segments. Random forest, being an ensemble of decision trees, can capture both linear and nonlinear relationships in the data due to the diversity of trees it comprises.

- **Homoscedasticity**: Decision trees do not make explicit assumptions about the homoscedasticity (constant variance) of errors. Similarly, random forest, being a combination of decision trees, is not directly affected by this assumption.

- **Normality of residuals**: Decision trees do not rely on the assumption of normality of residuals, and the random forest algorithm inherits this flexibility. However, if you're using random forest as part of a broader analysis that assumes normality (for example, hypothesis testing), you should consider this aspect in your overall approach.

- **Feature scaling**: Random forest is relatively insensitive to the scale of features. It doesn't require features to be standardized or normalized, unlike some other algorithms, such as gradient boosting or K-means clustering.

- **Multicollinearity**: Random forest can handle multicollinearity (high correlation between features) effectively as it considers only a subset of features at each node in each tree, reducing the potential impact of correlated features.

It's worth noting that while random forest is more robust and forgiving than individual decision trees, it is not entirely immune to the quality and characteristics of the data. Preprocessing, data cleaning, and understanding the data's domain-specific properties remain important steps in building accurate and reliable random forest models.

Common pitfalls

While random forest is a powerful algorithm, it's important to be aware of potential pitfalls:

- **Overfitting with too many trees**: Although random forest reduces overfitting, using an excessive number of trees can still lead to unnecessary computational complexity

- **Bias toward dominant classes**: In imbalanced datasets, random forest might favor the majority class due to its inherent averaging mechanism

- **Computation and memory**: Training a large random forest can be computationally expensive and memory-intensive

- **Feature selection**: While random forest provides feature importance, it might not always identify the optimal subset of features for a specific problem

Implementation example

Here's a basic example of how to implement a random forest classifier using Python and the popular scikit-learn library:

```
# Import necessary libraries and prep dataset
import numpy as np
from sklearn.model_selection import train_test_split
from sklearn.datasets import load_iris
from sklearn.ensemble import RandomForestClassifier
from sklearn.metrics import accuracy_score
iris = load_iris()
X = iris.data
y = iris.target
X_train, X_test, y_train, y_test = train_test_split(X, y, test_
size=0.2, random_state=42)


# Initialize the Random Forest classifier and compute accuracy
```

```
random_forest = RandomForestClassifier(n_estimators=100, random_
state=42)
random_forest.fit(X_train, y_train)
y_pred = random_forest.predict(X_test)
accuracy = accuracy_score(y_test, y_pred)
print(f'Accuracy: {accuracy:.2f}')
```

In this example, we have used the popular Iris dataset for simplicity. Here's what the code does:

- Again, we use the popular Iris classification dataset.

- The dataset is split into training (80% of the data) and testing (20% of the data) sets to evaluate the model's performance on unseen data.

- A random forest classifier, `RandomForestClassifier`, which is an example of an ensemble learning method, is initialized with 100 trees and then trained on the training data. Additionally, the model was set with a random seed (`random_state=42`), ensuring reproducibility.

- After training, the model's performance is evaluated by predicting the test set's class labels and subsequently calculating and printing the accuracy of these predictions compared to the true test set labels.

Assessment

How does a random forest work?

Answer

A random forest is an ensemble of decision trees. It trains multiple trees on different subsets of the data and combines their predictions through majority voting or averaging.

Assessment

What is the role of randomness in a random forest?

Answer

Randomness is introduced through bootstrap sampling of data and feature subsampling during tree construction. This helps in reducing overfitting and promoting diversity among trees.

Assessment

What are the advantages of using random forest?

Answer

Random forests are robust to overfitting, handle high-dimensional data well, provide feature importance scores, and can handle both classification and regression tasks.

Extreme Gradient Boosting (XGBoost)

XGBoost is a powerful and highly efficient gradient-boosting algorithm that's designed to tackle a wide range of machine learning problems. Like random forest, it can be used for both regression and classification. It has gained significant popularity due to its exceptional performance in predictive modeling competitions and real-world applications. XGBoost is particularly effective in handling structured/tabular data and is known for its robustness, scalability, and ability to capture complex patterns.

How it works

XGBoost builds an ensemble of weak predictive models (typically decision trees) sequentially, where each subsequent model tries to correct the errors made by the previous ones. The core principles of XGBoost are as follows:

- **Gradient boosting**: XGBoost employs gradient boosting, which involves minimizing a loss function by iteratively adding new models to the ensemble
- **Regularization**: XGBoost incorporates L1 (lasso) and L2 (ridge) regularization terms into the loss function to control overfitting
- **Feature importance**: XGBoost provides insights into feature importance, allowing you to understand the contribution of each feature to the model's predictions
- **Cross-validation**: XGBoost supports k-fold cross-validation to evaluate and optimize model performance

Assumptions

Similar to random forest, XGBoost is an ensemble learning algorithm based on decision trees and has fewer assumptions compared to traditional linear models. Thus, the same practical considerations mentioned for Random Forest should also be considered here.

Clarifying boosting versus bagging

XGBoost, as its name implies, relies on gradient boosting, or simply boosting. **Boosting** is an iterative technique that sequentially builds a strong model by combining multiple weak models. The idea is to focus on the examples that the current set of models is struggling with and assign them higher weights, effectively "boosting" their importance. The weak models are trained sequentially, and each new model gives more weight to the misclassified examples from the previous models.

However, in **bagging (bootstrap aggregating)**, multiple instances of the same algorithm are trained on different subsets of the training data, obtained by random sampling with replacement. The final prediction is typically an average or majority vote of the predictions from individual models. Random forest is a well-known ensemble method that uses bagging with decision trees.

So, the differences between bagging and boosting are as follows:

- **Combination of models**: Bagging involves training multiple base models independently and then aggregating their predictions. Boosting trains a sequence of models iteratively, where each new model focuses on correcting the errors of the previous models.

- **Training approach**: Bagging reduces variance by averaging predictions from diverse models. Boosting reduces both bias and variance by iteratively refining the model's performance.

- **Weight assignment**: Bagging assigns equal weights to all training examples. Boosting assigns higher weights to misclassified examples, focusing more on difficult instances.

- **Sequential versus parallel**: Bagging trains base models in parallel as they are independent of each other. Boosting trains models sequentially, where each new model depends on the performance of the previous models.

- **Performance**: Boosting often results in higher accuracy but may be more prone to overfitting if not controlled properly. Bagging tends to have lower variance and may be less prone to overfitting.

In summary, while both bagging and boosting aim to improve ensemble model performance, they differ in their approach to combining models and how they handle training instances. The choice between bagging and boosting depends on the nature of the problem, the available data, and the desired trade-off between bias and variance.

Common pitfalls

While XGBoost is a powerful algorithm, there are some considerations to keep in mind:

- **Hyperparameter tuning**: The performance of XGBoost can be sensitive to hyperparameters. Careful tuning is essential for optimal results

- **Overfitting**: Despite regularization, overfitting can still occur, especially when using a large number of boosting rounds

- **Computational complexity**: Complex models or large datasets can lead to increased computational demands and longer training times

- **Interpretability**: While XGBoost provides feature importance, its complex nature may make model interpretation challenging

Assessment

How does XGBoost work? What are the advantages of using XGBoost?

Answer

XGBoost sequentially adds new models to the ensemble, aiming to correct errors made by the previous models through gradient boosting. XGBoost offers high performance, flexibility, feature importance insights, regularization, and handling of imbalanced classes.

Assessment

What are the differences between bagging and boosting?

Answer

Bagging and boosting are both ensemble methods in machine learning that combine multiple models to improve performance. Bagging trains multiple base models independently and aggregates their predictions, typically reducing variance by averaging diverse model predictions. On the other hand, boosting trains models sequentially, where each subsequent model corrects the errors of its predecessors, aiming to reduce both bias and variance.

While bagging assigns equal weights to all training examples, boosting prioritizes misclassified instances by giving them higher weights. This means that boosting focuses more on challenging cases. Bagging's models are trained in parallel since they operate independently, whereas boosting requires a sequential approach because each model builds upon the performance of the previous ones. In terms of performance, boosting often achieves higher accuracy but can be more susceptible to overfitting if not managed carefully, whereas bagging generally offers more stability and is less prone to overfitting.

Getting started with unsupervised machine learning

Unsupervised machine learning is a fascinating branch of artificial intelligence that focuses on discovering patterns, relationships, and structures within data without explicit guidance from labeled outcomes. Unlike supervised learning, where models are trained with labeled data to make predictions, unsupervised learning aims to explore the inherent information present in the data itself. This type of learning is particularly valuable for uncovering hidden insights, finding clusters, reducing dimensionality, and revealing underlying representations. Clustering is a common use case for unsupervised learning.

Clustering refers to grouping data points into distinct subsets or "clusters" based on similarities in their features without using pre-labeled data as a guide. Imagine that you have a scatter plot of data points and want to color-code groups of points that seem to cluster together; this is essentially what clustering algorithms do but in potentially multi-dimensional spaces. The goal is to ensure that data points in the same cluster are more alike to each other than those in different clusters.

For businesses, clustering has numerous applications: customer segmentation for targeted marketing, organizing large sets of documents or news articles into cohesive topics, detecting abnormal patterns or anomalies in data, and even helping retailers optimize product placements in stores based on purchasing behaviors. Uncovering these natural groupings allows businesses to gain insights, enhance decision-making, and tailor strategies to specific audience segments.

In this section, we will delve into the foundational concepts of unsupervised learning, including its key algorithms, applications, challenges, and interview questions, shedding light on how it empowers us to extract meaningful knowledge from unannotated data. First, we will look at some common clustering algorithms and finish with how to evaluate the clusters produced by the algorithms.

K-means

K-means clustering is a fundamental unsupervised learning algorithm that's designed to partition data into distinct groups, or clusters, based on similarities between data points. It is widely used for pattern recognition, segmentation, and understanding the underlying structure within datasets. K-means is intuitive, computationally efficient, and can provide valuable insights into the inherent grouping of data.

How it works

K-means works by iteratively assigning data points to clusters and updating cluster centroids to minimize the sum of squared distances between points and their respective centroids. The key steps involved are as follows:

1. **Initialization**: Randomly select initial cluster centroids
2. **Assignment**: Assign each data point to the nearest centroid
3. **Update**: Recalculate centroids based on the mean of data points in each cluster
4. **Repeat**: Iterate between assignment and update until convergence or a specified number of iterations

Assumptions

While K-means is relatively simple and effective, it does make certain assumptions about the data and the structure of clusters. These assumptions can impact the algorithm's performance and the quality of the resulting clusters. Here are the key assumptions of K-means:

- **Cluster shape and size**: K-means assumes that clusters are spherical and have roughly equal sizes. In other words, it assumes that the clusters have similar diameters and contain roughly the same number of data points.

- **Equal variance**: K-means assumes that the variance (spread) of the data points within each cluster is roughly the same. This assumption is important because K-means uses the mean as the center of a cluster, and equal variance helps in determining the "average" distance of data points from the center.

- **Features' influence**: K-means treats all features equally and assumes that they have a similar influence on the clustering process. This can be problematic if some features are more relevant or important than others.

- **Independence of clusters**: K-means assumes that the clusters are independent and non-overlapping. In reality, data points may belong to multiple clusters or exhibit complex patterns that K-means cannot capture.

- **Globular clusters**: K-means works well for clusters that are roughly globular in shape. If clusters have irregular shapes, elongated structures, or densities, K-means may struggle to accurately capture these patterns.

- **Pre-defined number of clusters**: K-means requires you to specify the number of clusters (k) in advance. This can be a challenge if the true number of clusters is not known or if the data doesn't naturally divide into distinct clusters.

- **Similar density clusters**: K-means assumes that the clusters have similar densities. If some clusters are denser than others, K-means may struggle to correctly assign data points.

- **Feature scaling**: Like most other cluster algorithms, it is required to scale the features to ensure one does not influence the model more than others.

Common pitfalls

K-means has some considerations and challenges:

- **Number of clusters**: Choosing the optimal number of clusters (k) can be subjective and impact the results

- **Sensitive to initialization**: K-means' performance can vary based on the initial centroids

- **Cluster shape and density**: K-means assumes clusters are spherical and equally sized, which might not always align with the data

- **Outliers**: Outliers can significantly influence cluster centroids and affect results

Implementation example

Here's a simple example of how to implement K-means clustering using Python and the scikit-learn library:

```
# Import necessary libraries and prep data with 4 clusters
import matplotlib.pyplot as plt
from sklearn.datasets import make_blobs
from sklearn.cluster import KMeans
```

```
X, _ = make_blobs(n_samples=300, centers=4, cluster_std=0.60, random_
state=0)

# Initialize K-Means with 4 clusters and plot cluster centers
kmeans = KMeans(n_clusters=4)
labels = kmeans.fit_predict(X)
cluster_centers = kmeans.cluster_centers_
plt.scatter(X[:, 0], X[:, 1], c=labels, cmap='viridis')
plt.scatter(cluster_centers[:, 0], cluster_centers[:, 1], s=300,
c='red')
plt.xlabel('Feature 1')
plt.ylabel('Feature 2')
plt.title('K-Means Clustering')
plt.show()
```

Let's take a closer look at this example:

- Necessary libraries such as Matplotlib and relevant functions from scikit-learn are imported, and we create our own dataset (that is, a synthetic dataset), with four distinct clusters generated using the make_blobs function, producing 300 samples. In this example, we are making our dataset show how to use the K-means model.

- The K-means clustering algorithm is initialized to partition the data into four clusters.

- The K-means algorithm is fitted to the synthetic data, assigning each data point to one of the four clusters. The centers of these clusters are then determined.

- Using Matplotlib, the data points are visualized on a scatter plot, color-coded based on their assigned clusters. The centers of these clusters are also plotted in red, and the resulting plot showcases how the K-means algorithm has grouped the data.

Assessment

What's the objective of the K-means clustering algorithm?

Answer

K-means aims to partition data points into clusters by minimizing the sum of squared distances between each data point and the centroid of its assigned cluster.

Assessment

How does K-means initialize cluster centroids?

Answer

K-means can use strategies such as random initialization, k-means++, or custom initialization to determine the initial locations of cluster centroids.

Density-based spatial clustering of applications with noise (DBSCAN)

DBSCAN is a powerful unsupervised learning algorithm that excels at identifying clusters of arbitrary shapes in data. Unlike K-means, which assumes spherical clusters of equal size, DBSCAN discovers clusters based on the density of data points in the feature space and is particularly useful when dealing with noisy data and clusters of varying sizes and shapes.

In this chapter, we will delve into the intricacies of DBSCAN, its principles, advantages, limitations, implementation in Python, and real-world applications.

How it works

DBSCAN identifies clusters by considering two main parameters: the radius (**epsilon**) that defines the neighborhood of a data point and the minimum number of data points (`min_samples`) required to form a dense region. The algorithm operates as follows:

- **Core points**: A data point is considered a core point if there are at least `min_samples` data points within its epsilon neighborhood

- **Border points**: A data point is a border point if it is within the epsilon neighborhood of a core point but does not have enough neighbors to be considered a core point itself

- **Noise points**: Data points that are not core or border points are classified as noise points

DBSCAN starts by selecting an arbitrary data point, expanding its neighborhood, and recursively growing a cluster. This process continues until no more data points can be added to the cluster, at which point a new cluster is formed. This process is then repeated until all data points have been classified into clusters or marked as noise.

DBSCAN offers several advantages:

- **Cluster shape**: DBSCAN can identify clusters of arbitrary shapes, making it suitable for complex datasets

- **Noise handling**: DBSCAN can effectively handle noisy data and classify outliers as noise points

- **Cluster size**: DBSCAN can discover clusters of varying sizes and densities within the same dataset

- **Parameter robustness**: DBSCAN requires minimal parameter tuning compared to methods that require specifying the number of clusters in advance

Assumptions

DBSCAN does not make specific assumptions about the shape of clusters like some other clustering algorithms (such as K-means), but it does have certain assumptions and characteristics:

- **Density-based clusters**: The primary assumption of DBSCAN is that clusters are areas of higher density separated by areas of lower density. Points within a cluster are densely packed, and there are regions with lower point densities separating different clusters.

- **Density reachability**: DBSCAN uses the concept of "density reachability." A data point is considered to be density-reachable from another point if it lies within a specified distance (epsilon, ε) from the other point and the number of points within that distance exceeds a predefined threshold (MinPts).

- **Core points**: Core points are data points that have at least MinPts data points within a distance ε from them. These points are at the center of clusters.

- **Border points**: Border points are not core points themselves but are within the ε-distance of a core point. They may belong to a cluster but are not considered the central point of a cluster.

- **Noise (outliers)**: Data points that do not meet the criteria to be core points or border points are considered noise or outliers. They do not belong to any cluster.

- **Cluster connectivity**: DBSCAN forms clusters by connecting core points that are density-reachable from each other. This means that a chain of core points can be used to connect different parts of the same cluster, even if they are not directly density-reachable.

- **Arbitrary shape clusters**: Unlike algorithms such as K-means, DBSCAN can identify clusters with arbitrary shapes and does not assume that clusters are spherical or elliptical.

- **Parameter sensitivity**: DBSCAN requires two main parameters: ε (epsilon) and MinPts. The choice of these parameters can impact the results, and finding appropriate values can sometimes require experimentation and domain knowledge.

Common pitfalls

There are several common pitfalls that data scientists may experience when working with DBSCAN:

- **Choosing incorrect parameters**: DBSCAN requires two critical parameters: epsilon (ε) and MinPts. Epsilon determines the maximum distance between two points for one to be considered a neighbor of the other, and MinPts specifies the minimum number of points within ε to form a core point. Choosing inappropriate values for these parameters can lead to undesirable results, such as overfitting, underfitting, or identifying noise as clusters.

- **Sensitive to data scaling**: DBSCAN's density-based nature makes it sensitive to the scaling of features. When features have significantly different scales, the choice of epsilon might not work well for all dimensions equally. Standardizing or normalizing the data can help mitigate this issue.

- **Noise interpretation**: DBSCAN can identify noise as separate clusters or classify noise as an outlier category. The interpretation of these noise points depends on the context of the data and the problem. Misinterpreting noise as actual clusters can lead to misleading insights.

- **Uneven density clusters**: DBSCAN may struggle with clusters of varying densities. If the density within a cluster is not consistent, setting a global epsilon might not work well for all parts of the dataset. In such cases, using other clustering algorithms or considering different density parameters for different areas could be more appropriate.

- **High-dimensional data**: DBSCAN's effectiveness can diminish in high-dimensional spaces due to the "curse of dimensionality." As the number of dimensions increases, the distance between points becomes less meaningful, potentially leading to sparser clusters or identifying most points as noise. Dimensionality reduction techniques or considering other clustering methods might be necessary for high-dimensional data.

- **Outliers identification**: DBSCAN can be sensitive to outliers, classifying them as noise or forming small clusters around them. Handling outliers requires a clear understanding of the problem and the ability to distinguish between genuine clusters and noise.

- **Cluster shape assumption**: While DBSCAN is effective at identifying clusters of varying shapes, it might struggle with clusters with varying densities or clusters embedded within other clusters. In such cases, alternative clustering algorithms such as hierarchical clustering might be more suitable.

Implementation example

Here's a basic example of how to implement DBSCAN using Python and the DBSCAN module:

```
# Import the needed libraries and prep the dataset
from sklearn.cluster import DBSCAN
from sklearn.datasets import make_blobs
import matplotlib.pyplot as plt
X, _ = make_blobs(n_samples=300, centers=3, cluster_std=0.6, random_
state=0)

# Create DBSCAN model, fit it, and plot clusters
dbscan = DBSCAN(eps=0.5, min_samples=5)
labels = dbscan.fit_predict(X)
plt.scatter(X[:, 0], X[:, 1], c=labels, cmap='viridis', s=50)
plt.xlabel('Feature 1')
plt.ylabel('Feature 2')
plt.title('DBSCAN Clustering')
plt.show()
```

In this example, the code does the following:

- Generates synthetic data using the `make_blobs` function from `sklearn.datasets`.

- Creates a `DBSCAN` instance by specifying the `eps` (epsilon) parameter, which controls the maximum distance between two data points for them to be considered part of the same cluster.

- Specifies the `min_samples` parameter, which defines the minimum number of neighboring points required to form a core point.

- Fits the model to the data using the `fit_predict` method, which returns cluster labels for each data point. Finally, we visualize the clusters using a scatter plot.

Other clustering algorithms

There are tons of clustering algorithms that you may encounter in the data science world. Choosing the correct model is all about understanding the unique scenario and business assumptions associated with your problem. While reviewing every unsupervised model is beyond the scope of this chapter, here are some other models you may want to consider:

- **Hierarchical clustering** is a clustering technique that builds a hierarchy of clusters by recursively dividing or merging data points based on a similarity measure. Unlike other clustering methods that produce a single partitioning of the data, hierarchical clustering creates a tree-like structure of clusters, known as a dendrogram. This dendrogram provides insights into the hierarchical relationships between data points and clusters.

- **Spectral clustering** transforms the data into a lower-dimensional space using the Laplacian graph and then performs clustering in this transformed space. It's especially useful for clustering data with complex structures and is not limited by the shape of clusters.

- **Ordering Points To Identify Clustering Structure** (**OPTICS**) is a density-based clustering algorithm similar to DBSCAN. It creates an ordering of data points based on their density-connectedness. Unlike DBSCAN, it produces a reachability plot, which helps in visualizing varying densities and clusters of different sizes.

- **Fuzzy c-means** (**FCM**) is a clustering algorithm that extends the traditional K-means algorithm by allowing data points to belong to multiple clusters with varying degrees of membership. Unlike K-means, where each data point belongs to a single cluster exclusively, FCM assigns each data point a membership value for each cluster, representing the degree of belongingness to that cluster. This makes FCM a fuzzy clustering algorithm, where points can have partial membership in multiple clusters.

Evaluating clusters

Now that we have covered a few clustering algorithms, the following are some common methods for evaluating clustering algorithms:

- **Silhouette score**: This quantifies how similar an object is to its cluster (cohesion) compared to other clusters (separation). It ranges from -1 to 1, where higher values indicate better-defined clusters.

- **Elbow method**: This is a graphical representation of the eigenvalues (variance) of each principal component or factor in a dataset. In the context of clustering, it is used to understand how much variance is explained by each cluster as the number of clusters increases. In what is called the scree plot, the X-axis represents the number of clusters, and the Y-axis represents the sum of squared distances (inertia). The plot typically resembles a curve that decreases sharply at first and then starts to level off. The "elbow" point, where the rate of decrease changes, indicates the optimal number of clusters. The elbow method helps identify a point where adding more clusters does not significantly improve the model's fit to the data, striking a balance between minimizing intra-cluster distance and avoiding excessive model complexity.

- **Adjusted rand index (ARI)**: This measures the similarity between the true class assignments and the predicted clusters, adjusting for chance.

- **Normalized mutual information (NMI)**: This quantifies the amount of information that's shared between true class assignments and predicted clusters, normalized to account for cluster size.

Assessment

How does DBSCAN work?

Answer

DBSCAN clusters data points based on their density. It defines dense regions as clusters and identifies outliers as noise points. Points within a specified distance (epsilon) and a minimum number of neighbors (`min_samples`) are considered part of the same cluster.

Assessment

What types of clusters can DBSCAN identify?

Answer

DBSCAN can identify clusters of varying shapes, including dense clusters, sparse clusters, and clusters separated by areas of lower density.

Assessment

How does DBSCAN handle noise and outliers?

Answer

DBSCAN can identify noise points as data points that do not belong to any cluster. Outliers that are isolated from dense regions are considered noise, while inlier points close to dense clusters are included in those clusters.

Summarizing other notable machine learning models

In the dynamic landscape of machine learning, a plethora of models cater to diverse data and problem domains. In this section, we will highlight other notable models, each offering unique capabilities and addressing specific challenges. From text processing to survival analysis, we'll explore a spectrum of models that expand the horizons of machine learning applications.

So, let's take a look:

- **Generalized additive models (GAMs)**: GAMs extend linear regression by accommodating nonlinear relationships between variables. By employing smooth functions, GAMs offer a flexible framework to capture complex interactions and patterns in data, making them valuable tools for various domains, including environmental science, economics, and healthcare.

- **Naïve Bayes**: This is a probabilistic classifier grounded in Bayes' theorem. Despite its simplicity, Naive Bayes excels in text classification, spam filtering, and sentiment analysis. Its efficiency in handling high-dimensional datasets and quick training make it a go-to choice for many text-based tasks.

- **Support vector machines (SVMs)**: These are versatile algorithms renowned for their ability to learn both linear and nonlinear boundaries between classes. In the realm of classification, SVMs provide high accuracy and robustness. Linear SVMs excel in scenarios with linear separability, while kernel methods enable SVMs to tackle complex decision boundaries in non-linear datasets.

- **Market basket analysis**: Market basket analysis focuses on discovering associations between items that are frequently purchased together. Widely used in retail, it reveals patterns that drive product recommendations and marketing strategies. **Apriori algorithm** and FP-growth are notable techniques for extracting frequent itemsets.

- **Survival analysis**: This analysis is used to analyze time-to-event data, such as customer churn, medical prognosis, or failure prediction. Employing hazard functions and Kaplan-Meier curves, this model assesses the probability of an event occurring within a given time frame.

- **Natural language processing** (**NLP**): NPL tasks encompass a vast range of tasks, including sentiment analysis, named entity recognition, machine translation, and question answering. Advanced models such as transformer-based architectures, such as BERT and GPT, have revolutionized NLP tasks by learning contextual representations. Examples of NLP tasks include sentiment analysis, text classification, **named entity recognition** (**NER**), text generation, text summarization, speech recognition, **text-to-speech** (**TTS**), and semantic labeling to name a few!

- **Anomaly detection models**: Anomaly detection is crucial for spotting outliers and identifying unusual patterns that deviate from expected behavior. Models such as isolation forest, one-class SVM, **local outlier factor** (**LOF**), and autoencoders excel in uncovering anomalies in fraud detection, network security, and fault diagnosis.

- **Recommender systems**: Recommender systems predict user preferences and recommend items or content of interest. Collaborative filtering, content-based filtering, and hybrid models combine user behavior and item attributes to provide personalized recommendations. **Matrix factorization** (**NMF**), **alternating least squares** (**ALS**), user-based filtering, and content-based filtering are prominent techniques that are employed in this domain.

Understanding the bias-variance trade-off

In the journey of building machine learning models, understanding how well they perform on unseen data is paramount. Evaluating a model's performance provides insights into its effectiveness, generalization capabilities, and potential areas for improvement. In this section, we delve into the critical process of using test sets to assess model performance comprehensively.

Model evaluation is a crucial step in the machine learning pipeline that validates the utility of a model in real-world scenarios. It gauges how well the model's predictions align with actual outcomes, ensuring that the model can make accurate and reliable decisions beyond the training data. When assessing a model's performance, it's essential to consider two key aspects: bias and variance.

Bias refers to the error due to overly simplistic assumptions in the learning algorithm, leading to an underfit model that misses relevant relationships. On the other hand, **variance** arises when a model is excessively complex and captures noise in the training data, resulting in an overfit model that doesn't generalize well to new data:

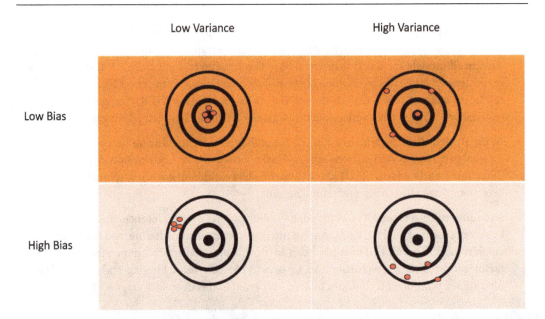

Figure 10.4: Depiction of bias versus variance

Striking the right balance between bias and variance is a delicate challenge. Increasing model complexity tends to reduce bias but can increase variance, while reducing complexity can lower variance but may increase bias. Achieving an optimal trade-off between bias and variance is crucial to developing models that can perform well on both training and test data.

Model complexity refers to the intricacy and flexibility of a machine learning model in capturing relationships within the data. A more complex model can fit the training data more closely, potentially capturing intricate patterns and noise. However, this increased complexity can also lead to **overfitting**, where the model becomes highly tailored to the training data and struggles to generalize to new, unseen data. On the other hand, a less complex model might not capture all the nuances of the data, leading to **underfitting**, where it fails to capture even the basic relationships present in the data.

Assessment

What is the bias-variance trade-off in machine learning?

Answer

The bias-variance trade-off refers to the balance between a model's ability to fit the training data well (low bias) and its ability to generalize to new, unseen data (low variance). Increasing model complexity can reduce bias but increase variance, and vice versa.

Assessment

How does underfitting relate to the bias-variance trade-off?

Answer

Underfitting occurs when a model is too simple to capture the underlying patterns in the data, leading to high bias and poor performance on both training and test data.

Assessment

How does overfitting relate to the bias-variance trade-off?

Answer

Overfitting happens when a model is too complex and fits the training data noise, resulting in low bias but high variance. This can lead to excellent performance on training data but poor generalization to test data.

Tuning with hyperparameters

Hyperparameter tuning is the process of systematically searching for and selecting the optimal values for the hyperparameters of a machine learning model. Unlike model parameters, which are learned from data during training, hyperparameters are determined by the practitioner and define characteristics such as the complexity of the model, the learning rate, regularization strength, and more. The goal of hyperparameter tuning is to identify the hyperparameter values that lead to the best possible model performance on unseen data.

Hyperparameter tuning involves experimenting with different values for each hyperparameter and evaluating the model's performance using appropriate evaluation metrics, often on a validation set. This process can be guided by different strategies, such as grid search, random search, or more advanced techniques such as Bayesian optimization.

Grid search

Grid search is a systematic approach to hyperparameter tuning. It involves defining a grid of possible hyperparameter values and exhaustively searching through all combinations. Grid search evaluates each combination using a predefined evaluation metric and identifies the configuration that yields the best performance.

While grid search guarantees thorough exploration of the hyperparameter space, it can be computationally expensive, especially when dealing with a large number of hyperparameters or a wide range of values.

Random search

Random search takes a different approach by randomly sampling hyperparameter combinations from predefined ranges. This stochastic method explores a broader range of hyperparameter values in fewer iterations compared to grid search. While it might not guarantee exhaustive coverage, random search has shown to be effective in discovering good hyperparameter configurations with less computational cost.

Bayesian optimization

Bayesian optimization leverages probabilistic models to efficiently navigate the hyperparameter space. It uses the information gained from previous evaluations to guide the selection of subsequent hyperparameter combinations. Bayesian optimization strikes a balance between exploration (trying new combinations) and exploitation (focusing on promising areas), making it highly efficient for hyperparameter tuning.

Assessment

What are hyperparameters in the context of machine learning?

Answer

Hyperparameters are parameters that are set before the learning process begins and influence a model's behavior and performance. They are not learned from data but are determined by the practitioner.

Assessment

How do hyperparameters impact model training?

Answer

Hyperparameters influence aspects such as model complexity, convergence speed, and regularization. Tuning hyperparameters can significantly impact a model's performance and generalization capabilities.

Assessment

What are common techniques for tuning hyperparameters?

Answer

Common techniques include grid search, random search, and more advanced methods such as Bayesian optimization. These methods systematically explore the hyperparameter space to find the best configuration for the model.

Summary

In our study of machine learning, we delved deeply into crucial concepts, obtaining significant insights. Our exploration spanned both supervised and unsupervised learning, equipping us with a diverse set of models.

In this chapter, we harnessed models ranging from linear and logistic regression to tree-based techniques such as random forests and XGBoost. These models have enabled us to capture intricate relationships and accurately estimate class probabilities. Additionally, our foray into clustering methods, including K-means, hierarchical clustering, and DBSCAN, has allowed us to master the art of extracting patterns from unlabeled data. Furthermore, our knowledge has been augmented with vital skills in hyperparameter tuning and model evaluation. We learned how to refine models using tools such as grid search and have come to understand key evaluation metrics, such as accuracy and precision.

As we gear up for data science interviews, this knowledge stands as a testament to our adaptability and problem-solving prowess. Beyond interviews, this understanding empowers us to tackle real-world data challenges and tailor models to meet diverse business needs. Our journey equips us to excel in interviews and make meaningful contributions to the dynamic world of data science.

In the next chapter, we will investigate deep learning concepts such as popular neural network architectures.

11

Building Networks with Deep Learning

In the previous chapter, we explored **machine learning** (**ML**) concepts, including common strengths, weaknesses, pitfalls, and various popular ML algorithms.

In this chapter, we will explore **artificial intelligence** (**AI**) as we dive into **deep learning** (**DL**) concepts. We will review important **neural network** (**NN**) fundamentals, components, tasks, and DL architectures that are most common in data science interviews. In doing so, we will unravel the mysteries of weights, biases, activation functions, and loss functions while mastering the art of gradient descent and backpropagation.

Along the way, we'll fine-tune our networks, delve into the magic of embeddings and **autoencoders** (**AEs**), and harness the transformative power of transformers. Plus, we'll unlock the secrets of **transfer learning** (**TL**), understand why NNs are often referred to as "black boxes," and explore common network architectures that have revolutionized industries and led the way for **generative AI** (**GenAI**) and **large language models** (**LLMs**) such as ChatGPT.

In this chapter, we will review the following topics:

- Introducing NNs and DL
- Weighing in on weights and biases
- Activating neurons with activation functions
- Unraveling backpropagation
- Using optimizers
- Understanding embeddings
- Listing common network architectures
- Introducing GenAI and LLMs

Introducing neural networks and deep learning

At its core, a **neural network** (also known as a neural net) is a computational model inspired by the structure and function of the human brain. It's designed to process information and make decisions in a manner akin to how our neurons work.

An NN consists of interconnected nodes, or artificial neurons, organized into layers. These layers typically include an input layer, one or more hidden layers, and an output layer, which you can see in *Figure 11.1*. Each connection between neurons is associated with a weight, which determines the strength of the connection, and an activation function, which defines the output of the neuron:

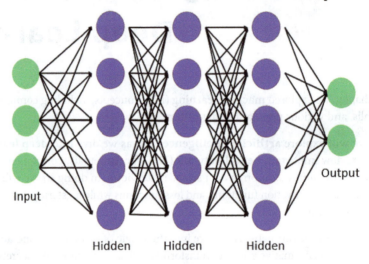

Figure 11.1: Basic NN diagram

Data passes from the input layer through the hidden layers until it reaches the final layer as an output. The preceding diagram shows two output nodes, but an NN can consist of one or even hundreds of output nodes. The number of output nodes is an important design decision when creating an NN. A data scientist must design the network to be effective with the problem they are working on. For example, an NN might only have one output node for a regression problem, while for a classification task, there may be an output node for each class.

In simpler terms, an NN takes input data, processes it through multiple hidden layers of interconnected neurons, and produces an output. This process of transforming input into output involves complex mathematical operations, but at its essence, NNs excel at learning patterns and making predictions from data.

Deep learning is a specific application of NNs in the ML field that focuses on training NNs with multiple hidden layers – hence the term "deep." While a standard NN might have just one or two hidden layers, DL models can have tens, hundreds, or even thousands of hidden layers. This depth allows them to learn intricate and hierarchical representations of data, making them particularly well suited for complex tasks such as image and speech recognition, **natural language processing (NLP)**, and more.

There are several benefits of using DL algorithms compared to their traditional ML counterparts:

- **Feature learning (FL)**: DL algorithms excel at automatically discovering features and patterns in data without explicit programming. They learn from vast amounts of data and adapt their internal representations to improve their performance on specific tasks. This ability to automatically extract features and make high-level abstractions from raw data is one of the key reasons DL has revolutionized fields such as **computer vision (CV)**, **natural language understanding (NLU)**, and **reinforcement learning (RL)**.

- **Complex data types**: DL excels at handling complex data types, such as images, audio, and **natural language (NL)** text. Traditional ML models may struggle to capture intricate patterns and structures present in these data types.

- **Scalability**: DL models can scale to handle large and complex datasets. With the increasing availability of powerful hardware (for example, GPUs and TPUs) and distributed computing, DL models can process massive amounts of data efficiently. This scalability is crucial in domains such as CV, where datasets can contain millions of images, or in training LLMs such as **Generative Pre-Trained Transformer 3 (GPT-3)**.

- **Applications**: DL models have achieved state-of-the-art performance in a wide range of applications, including image recognition, speech recognition, machine translation, and game-playing. Their ability to capture intricate patterns and representations allows them to outperform traditional ML models in many cases.

- **Transfer Learning**: DL models can leverage pre-trained NNs and transfer knowledge from one task to another. For example, pre-trained models such as **Bidirectional Encoder Representations from Transformers (BERT)**, which was originally designed for NLU, have been fine-tuned for various NLP tasks, demonstrating their adaptability. TL enables faster and more efficient training on new tasks with limited data, making DL practical for real-world scenarios where collecting large datasets may be expensive or time-consuming.

Assessment

What is the primary difference between a standard NN and a DL model, as mentioned in the text?

- A. DL models are inspired by the human brain, while standard NNs are not

- B. DL models have multiple hidden layers, sometimes even thousands, while standard NNs might have one or two

    C.   DL models do not use activation functions

    D.   Standard NNs can handle complex data types such as images, audio, and text, while DL models cannot

Answer

The correct answer is *B*. DL focuses on training NNs with multiple hidden layers, whereas a standard NN might have just one or two hidden layers.

Assessment

Which of the following benefits of DL is highlighted by its ability to learn from vast amounts of data and adapt its internal representations for specific tasks without explicit programming?

    A.   Scalability

    B.   Complex data types

    C.   Feature Learning (FL)

    D.   Applications

Answer

The correct answer is *C*. DL algorithms excel at "*automatically discovering features and patterns in data without explicit programming.*"

Weighing in on weights and biases

Weights and biases are some of the most important components of NNs. Their functionality within NN nodes complements each other, similar to how weights and biases fit linear regression models. Understanding weights and biases will help you understand how they transform an NN from a static structure into a dynamic learning system. Proficiency in initializing, updating, and optimizing these components is essential in the journey of training NNs effectively.

Introduction to weights

Weights are numerical values that are assigned to the connections between neurons. Each connection possesses a corresponding weight value, which dictates the strength of the influence one neuron has on another. During training, these weights are adjusted, enabling the network to capture patterns and relationships within the data it processes.

Initially set to random values, these weights are fine-tuned through techniques such as backpropagation and gradient descent, which we'll discuss later. This fine-tuning process is the core mechanism through which NNs learn and adapt to different tasks.

Introduction to biases

Biases serve as essential parameters in NNs, akin to constants that influence the behavior of individual neurons within a layer. They are added to the weighted sum of inputs to a neuron *before* the activation function is applied. Biases allow the network to account for variations and offsets in the input data, enhancing its adaptability and flexibility.

As with weights, biases are initialized with small values and updated during training. They play a crucial role in ensuring that the NN can effectively capture complex relationships within the data. For example, *Figure 11.2* demonstrates how your model's inputs, weights, and biases produce an output for a single node:

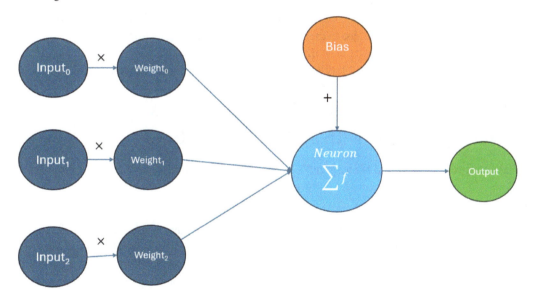

Figure 11.2: Weights and biases in a node of a simple network

In this example, we witness a "**forward pass**," which involves passing the input data to the NN where the weights and biases are used to produce an output. The process involves the following steps:

1. Each model input is multiplied by its respective weight.

2. The sum of the weight and input products is computed in the neuron.

3. The bias value is added to the weighted sum.

4. An activation function (more on this shortly) is applied to the remaining value.

5. The result is the model's output.

While this example goes over a single propagation within an incredibly basic, single-layer network, note that most DL models have tens, hundreds, and even thousands of hidden layers. In *Figure 11.1*, we saw an example of a simple NN with three hidden layers. The more complex the model, the more hidden layers are required, resulting in a deeper NN. In the next section, we will review activation functions, which aid our weights and bias in producing the model's output.

Assessment

Explain the role of biases in NNs and how they differ from weights.

Answer

Biases serve as essential parameters in NNs, acting as constants that influence the behavior of individual neurons within a layer. They are added to the weighted sum of inputs to a neuron before the activation function is applied. Biases allow the network to account for variations and offsets in the input data, enhancing its adaptability and flexibility. On the other hand, weights are numerical values that are assigned to connections between neurons that dictate the strength of influence one neuron has on another. During training, these weights are adjusted to capture patterns and relationships in the data.

Activating neurons with activation functions

We reviewed how weights and biases contribute to a model's predictions in the previous section. However, the fourth step in *Figure 11.2* involves something called an activation function. What is an activation function anyway?

In the intricate architecture of NNs, activation functions are the gears that infuse life and non-linearity into the system. **Activation functions** are mathematical functions that are applied to the output of each neuron, introducing non-linearity to the outputs. This is a key distinction between the application of weights and biases in linear regression. Let's explore the role and types of activation functions that breathe vitality into NNs.

At its core, **non-linearity** allows NNs to capture complex patterns in data that a linear approach would miss. Imagine trying to fit a straight line to data that twists and turns in various directions. A linear model would fail to capture the intricacies, but with non-linearity, a model can bend and adjust to these curves, making it more adaptable and accurate.

Within the intricate framework of NNs, activation functions are like the heartbeats that introduce this non-linearity. They are mathematical formulas that are applied to the output of each neuron, ensuring the outputs aren't just a straight-line prediction. Including non-linearity is a pivotal point that differentiates NNs from linear models such as linear regression.

Referring back to *Figure 11.2*, an activation function operates at the neuron. The inputs, multiplied by their weights plus the bias value, are all summed together and given as input to the activation function. The output of the activation function is based on this input. For example, the step activation function, which we will highlight again shortly, returns 1 if the input is greater than 0 and 0 for anything less than 0. This output may go on to the next become the input of the next neuron, and the process starts again.

Common activation functions

Now, let's look at some of the most common activation functions that you will encounter when building NNs. For each of the formulas in the list, we have the following:

- *e* represents the mathematical constant Euler's number (approximately 2.71828)
- *zi* is an element of the input vector, z
- The denominator, which is the sum of the exponential values of all elements in the input vector

Here is a list of formulas:

- **Step**: The output of the step function (also known as the Heaviside step function) is either 0 or 1. It says that if the value is 0 (or less than 0), then return 0. Otherwise, if it's anything greater than 0, return 1. Hence, the step function is a "strong function" because there's not much room for ambiguity:

$$H(x) = \begin{cases} 1 \, for \, x \geq 0 \\ 0 \, for \, x < 0 \end{cases}$$

- **Sigmoid**: The sigmoid activation function squashes the input values into the range of [0, 1]. It's often used in the output layer of binary classification tasks where the network needs to produce probabilities:

$$\theta(x) = \frac{1}{(1 + e^{-x})}$$

- **Hyperbolic tangent (tanh)**: Tanh is similar to sigmoid but squashes input values into the range of [-1, 1]. It's often used in hidden layers of NNs:

$$\tanh(x) = \frac{(e^x - e^{-x})}{(e^x + e^{-x})}$$

- **Rectified Linear Unit (ReLU)**: ReLU is one of the most popular activation functions. It replaces negative inputs with zero and passes positive inputs unchanged. ReLU has been highly effective in training **deep NNs (DNNs)**:

$$ReLU \, (x) = \max \, (0, x)$$

- **Leaky ReLU**: Leaky ReLU is a variant of ReLU that allows a small, nonzero gradient for negative inputs to avoid the "dying ReLU" problem, where neurons get stuck in a non-active state:

$$LeakyReLU(x) = \begin{cases} x, & if\, x \geq 0 \\ negative\, slope \times x, & otherwise \end{cases}$$

- **Softmax**: The softmax function ensures that the output probabilities sum to 1, making it suitable for multiclass classification tasks:

$$\theta(z)_i = \frac{e^{z_i}}{\sum_j^K e^{z_j}} for\, i = 1, ..., K$$

> **Note**
>
> Softmax is often used in the output layer of NNs for tasks such as image classification, NLP, and various other classification problems.

- **Linear**: You are already familiar with this function:

$$f(x) = ax + b$$

Choosing the right activation function

The choice of activation function depends on the problem at hand and the characteristics of the data. Here are some examples:

- Sigmoid and tanh are suitable for specific scenarios such as binary classification, where the output needs to be in a bounded range.

- Sigmoid is also used in multiple-label, multiple-class problems.

- ReLU and its variants are often preferred for DNNs due to their ability to mitigate the vanishing gradient problem, which can hinder training in deeper architectures. We will talk about the vanishing gradient problem later in this chapter.

- Softmax is suitable for multiclass classification problems (single label, multiple classes).

Experimentation and consideration of the activation functions' properties, such as range, are crucial in selecting the right one for your NN.

Assessment

What role do activation functions play in NNs, and why is non-linearity crucial for these systems?

Answer

Activation functions introduce non-linearity into NNs by being applied to the output of each neuron. This non-linearity ensures that the NN can capture and model complex relationships in the data, which a linear model might not be able to represent.

Non-linearity can be understood as the property where the outcome does not change in direct proportion to a change in any of the inputs. Without non-linearity, every layer of an NN would essentially be a linear transformation, and no matter how many layers are added, the final output would still be a linear function of the input. Therefore, activation functions are essential for NNs to learn from complex datasets.

Assessment

Take a look at the following three scenarios:

- A. An NN layer that needs to produce probabilities for binary classification
- B. The output layer of an NN that is meant for image classification with multiple categories
- C. A hidden layer in a DNN where the vanishing gradient problem could be an issue

Given these scenarios, pick the most appropriate activation function in terms of sigmoid, ReLU, softmax, and tanh.

Answer

Here are the answers:

- A. **Sigmoid**: The sigmoid activation function squashes input values into the range of [0, 1], making it suitable for producing probabilities, especially in binary classification tasks
- B. **Softmax**: The softmax function ensures that the output probabilities sum to 1, making it suitable for multiclass classification tasks such as image classification with multiple categories
- C. **ReLU**: ReLU and its variants are commonly used in hidden layers of DNNs due to their ability to mitigate the vanishing gradient problem, which can hinder training in deeper architectures

Unraveling backpropagation

At this point, you may be wondering why weights, biases, and activation functions are so special. After all, at this point, they probably seem not much different than parameters and hyperparameters in traditional ML models. However, understanding backpropagation will solidify your appreciation of how weights and biases work. This journey begins with a brief discussion of gradient descent.

Gradient descent

In short, **gradient descent** is a powerful optimization algorithm that's widely used in ML and DL to minimize a cost or loss function. It is the name that's given to the process of training a model on a task by first making a prediction with the model, measuring how good that prediction is, and then adjusting its weights slightly so that it will perform better next time. This process allows the model to gradually make better predictions over many iterations of training. It is used to train not only NNs but also other ML models, such as linear and logistic regression and **principal component analysis (PCA)**.

To adjust the weights to improve the model, the error gradient concerning each of the weights is computed. In essence, this means knowing how much each weight influenced the prediction error. To do this with NNs, we use the backpropagation algorithm.

What is backpropagation?

Backpropagation, also known as "backward propagation of errors," is a fundamental algorithm that's used to train **artificial NNs (ANNs)**. It uses the chain rule from calculus to compute gradients quickly and efficiently. The process was invented in the 1970s, but it wasn't until the 1980s from the work of Hinton and others that the algorithm was appreciated by the ML community. Just take a moment to appreciate that this simple algorithm allows NNs to be trained with a million+ weights.

The gradients point in the direction of the steepest ascent, and gradient descent takes steps in the opposite direction to minimize the loss. *Figure 11.3* displays a two-dimensional gradient descent graph where a given parameter, *p*, is minimized to the global loss minimum:

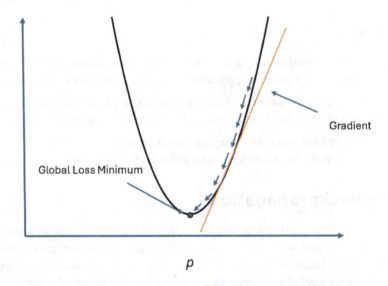

Figure 11.3: Optimizing for the global loss minimum

If you were to flip the parabola in *Figure 11.3* such that the opening faced down and the apex faced up, the point in the middle would represent the global maximum instead. Gradient descent typically involves finding either a maximum or minimum value of a parameter.

Loss functions

Loss functions, also known as cost functions or objective functions, serve as critical guides in training models, helping them understand how well they are performing on a given task. These functions quantify the disparity between predicted values and true target values, providing a measure of error.

Let's quickly review some loss function examples and their respective errors (you may recognize many of the error metrics from *Chapter 10*):

Loss Function	Error
Regression	Mean squared error (MSE), mean squared logarithmic error (MSLE), mean absolute error (MAE)
Binary classification	Binary cross-entropy, hinge loss, squared hinge loss
Multiclass classification	Multiclass cross-entropy, sparse multiclass cross-entropy

Figure 11.4: Loss functions

Gradient descent steps

The following are the basic steps that are taken in the backpropagation process:

1. **Forward pass**: This is what we saw earlier in *Figure 11.2*. During the forward pass, input data is fed into the NN, and it passes through each layer of neurons, including the input layer, hidden layers (if any), and the output layer. At each neuron, the weighted sum of inputs is computed, followed by the application of an activation function, which determines the neuron's output; this process continues through the network until it produces a final output.

2. **Calculate the error**: Once the network makes predictions, the next step is to calculate the error or loss between the predicted output and the actual target values. The choice of the error metric depends on the specific task; for example, MSE is common for regression, while cross-entropy is used for classification tasks.

3. **Backward pass (backpropagation)**: In this critical phase, the error is propagated backward through the network, layer by layer. The goal is to determine how much each parameter (weights and biases) contributed to the error. This is done by calculating the gradient of the error concerning each parameter using the chain rule from calculus.

4. **Update the parameters**: With the gradients in hand, the network updates its parameters (weights and biases) in the opposite direction of the gradient. This step aims to reduce the error by making small adjustments to the parameters. The size of these adjustments is controlled by a hyperparameter called the learning rate.

5. **Reiterate**: *Steps 1 to 4* are repeated iteratively for a specified number of **epochs** (times) or until the error converges to a minimum. During each iteration, the network refines its parameter values, attempting to minimize the error on the training data.

6. **Validation and testing**: After training, the NN's performance is evaluated on validation data to ensure it generalizes well to unseen examples. Testing is performed on a separate test dataset to assess the model's performance in real-world scenarios.

In short, forward propagation uses model inputs as signals, while backpropagation uses model errors as input signals. By constantly re-evaluating their performance and tweaking weights and biases, DL networks can self-correct their mistakes. In turn, DL models almost eliminate the lengthy hyperparameter tuning required in ML models.

The vanishing gradient problem

The **vanishing gradient problem** is a challenge that occurs during the training of DNNs, particularly those with many layers. It is characterized by diminishing gradient values as they are backpropagated from the output layer to the earlier layers during training. When gradients become too small, the network's weights and biases are updated very slowly or not at all, resulting in slow or halted learning.

There are several reasons why vanishing gradients may occur:

- **Chain rule and backpropagation**: During backpropagation, the gradients of the loss function concerning the parameters (weights and biases) in each layer are calculated using the chain rule. Gradients are propagated backward from the output layer to the input layer.

- **Activation functions**: In deep networks, non-linear activation functions such as sigmoid or tanh are often used. These functions squash their input values into a limited range, resulting in derivatives that are small when inputs are far from zero.

- **Cumulative effect**: As gradients are calculated layer by layer during backpropagation, the derivatives of the activation functions are multiplied together. If these derivatives are consistently small, the gradients can shrink exponentially as they move backward through the layers.

- **Weight initialization**: Initial weight values can also contribute to the vanishing gradient problem. If weights are initialized with very small values, the gradients in the early layers may become too small to drive effective updates.

Meanwhile, the **exploding gradient problem** is the counterpart of the vanishing gradient problem. Instead of gradients becoming excessively small, they become exceptionally large during backpropagation, leading to numerical instability during training. When gradients explode, they can cause weight updates that are so large that they overshoot the optimal parameter values and prevent the model from converging.

Here are some possible reasons why exploding gradients may occur:

- **Gradient magnification**: In deep networks, the gradients of the loss function concerning the parameters can amplify as they are calculated and propagated backward through the layers. This amplification occurs when the derivatives of activation functions are greater than one.

- **Weight initialization**: Poor choices of weight initialization, especially when initial weights are too large, can exacerbate the exploding gradient problem. If weights are initialized with values that are too large, gradients can explode during training.

The vanishing gradient problem can hinder the training of deep networks, especially **recurrent NNs (RNNs)**, which are special kinds of NNs that are often used when working with time series data, and networks with many layers (deep feedforward NNs or **convolutional NNs (CNNs)**). It often results in slow convergence, and the network may struggle to capture long-term dependencies in sequential data. The exploding gradient problem can lead to model instability, divergence during training, and numerical overflow issues.

To mitigate it, techniques such as **gradient clipping**, which is a technique that limits gradient values during training, and careful weight initialization are often employed. It works by setting a threshold value, and if the gradient exceeds this value, it is scaled down to keep it within a certain range. This prevents the weights from being updated too drastically, maintaining stability in the training process. There are two main types of gradient clipping: value clipping and norm clipping. In value clipping, each element of the gradient is clipped individually. If a gradient component is greater than the positive threshold, it is set to the threshold. Similarly, if a component is less than the negative threshold, it is set to the negative threshold. However, in norm clipping, instead of clipping each gradient value individually, the entire gradient vector is scaled down.

You may also explore one of the following initialization methods to avoid vanishing and exploding gradients:

- **Glorot or Xavier**: This is a technique that's used to initialize weights in such a way that the variance of the activations is the same across every layer, which helps prevent the gradient from exploding or vanishing. It is best used with tanh, sigmoid, and softmax activation functions.

- **He**: Similar to the Glorot method, the He weight initialization method focuses on initializing weights in such a way that the variance of the activations is the same across every layer. However, the methods differ in the way they calculate the variance of weights. This method is best used with ReLU and its variants.

Both the vanishing and exploding gradient problems are critical challenges in training DNNs, and addressing them is essential for the successful convergence of DL models. Techniques such as using appropriate activation functions, careful weight initialization strategies, gradient clipping, and architectural modifications such as skip connections have been developed to alleviate these issues and enable deep networks to be trained effectively.

Assessment

Describe backpropagation and its relation to gradient descent and loss functions in the context of training NNs.

Answer

Backpropagation, also known as "backward propagation of errors," is a central algorithm that's used for training ANNs. It's the method through which these networks learn from their mistakes by adjusting their internal parameters, namely weights and biases, to enhance performance on specific tasks.

Backpropagation is closely related to gradient descent and loss functions in the following way:

- **Loss functions**: These are essential metrics that help models understand their performance by quantifying the difference between predicted values and the actual target values. The error or loss that's calculated using loss functions is a critical input to the backpropagation process.

- **Gradient descent**: Gradient descent is an optimization algorithm that is utilized in the iterative process of refining model parameters, such as weights and biases, to find the best values that minimize the cost or loss function. Backpropagation aids in determining how much each parameter (weights and biases) contributed to the error by calculating the gradient of the error concerning each parameter. This gradient is then used in the gradient descent algorithm to update the model parameters, guiding the model toward better performance.

Assessment

Which of the following statements is true regarding the vanishing and exploding gradient problems in DNNs?

A. The vanishing gradient problem results from gradients becoming excessively large during backpropagation

B. The exploding gradient problem can cause weight updates that are so large that they prevent the model from converging

C. Activation functions such as ReLU are the primary reasons for the vanishing gradient problem

Answer

- *A* is false – the vanishing gradient problem is characterized by diminishing gradient values, not increasing ones

- *B* is true – when gradients explode, weight updates can become so large that they prevent the model from converging

- *C* is false – the vanishing gradient problem often arises due to activation functions such as sigmoid or tanh, not ReLU

Using optimizers

At the heart of DL lies the optimization problem: finding the best set of model parameters (weights and biases) that minimize a chosen loss function. Optimization algorithms play a pivotal role in this journey by iteratively adjusting these parameters to reduce errors between predictions and actual target values.

Optimization is a fundamental concept in mathematics that refers to the process of finding the best or most favorable solution among a set of possible solutions. In the context of ML and DL, optimization is used to adjust model parameters to minimize a cost, objective, or loss function (all used interchangeably), leading to improved model performance. We have already covered that the gradient descent algorithm is used for optimization. However, there are different versions of the algorithm, and when constructing your NN, you can choose which of them to use.

Let's consider some key aspects of optimization:

- **Objective function**: Optimization involves an objective function, also known as a cost function or loss function, as mentioned earlier. This function quantifies the difference between the predicted values of a model and the actual target values. The goal is to minimize (or maximize, in some cases) this function.

- **Local minimum**: A local minimum is a point in the solution space where the objective function has a lower value than at all nearby points but may not necessarily be the absolute lowest point in the entire solution space. It's like a dip in a hilly landscape where you're at the lowest point around, but there might be deeper valleys elsewhere.

- **Global minimum**: A global minimum is the absolute lowest point in the entire solution space, where the objective function has its smallest value. It represents the best possible solution to the optimization problem. Finding the global minimum can be challenging, especially in complex, high-dimensional spaces.

The optimization process is always looking for the global minimum but can sometimes get stuck in a local minimum. Different versions of the gradient descent algorithm were developed with different approaches to find the global minimum, and we will talk about them in our next section.

Optimization algorithms

Optimization algorithms, such as gradient descent and its variants, are employed to navigate through the solution space to find either the global minimum or a satisfactory local minimum, depending on the problem. The choice of optimization algorithm, learning rate, and other hyperparameters can significantly impact the convergence of the optimization process and the quality of the solution found.

While gradient descent forms the bedrock, numerous advanced optimization algorithms have been developed to address its limitations and accelerate training. Some common ones include the following:

- **Stochastic gradient descent (SGD)**: An extension of gradient descent that computes gradients and updates parameters using mini-batches of training data, making it more computationally efficient.

- **Adaptive Moment Estimation (Adam)**: An adaptive learning rate optimization algorithm that combines the advantages of momentum and **Root Mean Squared Propagation** (**RMSprop**). Adam adjusts the learning rate individually for each parameter.

- **RMSprop**: An optimization algorithm that adapts the learning rate for each parameter based on the magnitude of recent gradients.

- **Adaptive Gradient Algorithm (AdaGrad)**: An optimization algorithm that adjusts learning rates adaptively based on the historical gradient information for each parameter.

- **Adadelta**: A variant of AdaGrad that addresses its sensitivity to the initial learning rate.

- **Nadam**: A combination of **Nesterov Accelerated Gradient** (**NAG**) and Adam that offers improved convergence properties.

Choosing the right optimizer is as much an art as it is a science. The optimal choice often depends on the specific problem, dataset, and model architecture. Furthermore, understanding the interplay between learning rates, batch sizes, and optimization algorithms is crucial for efficient training and model convergence.

Optimizers are the captain's wheel that steers the ship of NN training. As we navigate through the intricacies of DL, mastering the art of optimization will empower us to train models that not only learn but excel in a wide range of tasks.

Network tuning

There are some common parameters that you should consider while improving model performance:

- **Epochs**: The number of "runs" or the number of times the NN trains on all the training data. One epoch means one complete pass through the entire training dataset. Two epochs represent two runs over the training data. We mentioned epochs earlier when we considered gradient descent. While increasing this value adds complexity to the model, it isn't the most effective way of improving results.

- **Batch size**: The number of samples fed to the model at a time. The model will update its weights after processing a batch. If the batch size is too small, it may lead to noisy gradients slowing the optimization process. However, if it is too large, it requires more computational resources, which may make training slower and more expensive.

- **Hidden Layers (n_hidden)**: The number of hidden layers. The more layers, the more complex the model (for more complex tasks). It will also take more time to run, so reducing the number of epochs a little may be helpful when increasing this parameter. Note that the hidden layers begin after the input layer, so they don't include it.

- **Dropout**: The drop parameter randomly drops some values propagated in the network during training with X% (where X = dropout rate). Some of the input values to the NN are randomly set to zero. This serves as a form of regularization as it forces the network to learn redundant patterns for better generalization. This is because each neuron becomes more capable since it cannot fully rely on its neighbors.

- **Optimizers**: The specific algorithm that's used to update the weights during model training. Examples include gradient descent, SGD, RMSprop, and Adam.

- **Learning rate**: This quantifies how quickly the optimizer converges. The larger the learning rate, the more likely it may "overstep" the optimal values. Smaller learning rates are more precise but take longer to train.

- **Regularization**: It's ideal to use regularization where there is overfitting in the model. Examples of regularization include L1 regularization such as the Lasso technique or L2 regularization such as the Ridge technique.

- **Batch normalization**: This increases training speed and accuracy because it helps prevent activations from becoming either too small or vanishing or too big or exploding.

Assessment

Which of the following statements best describes the relationship between a local minimum and a global minimum in the context of optimization?

A. A local minimum is always higher in value than a global minimum

B. A local minimum is the absolute lowest point in the solution space, while a global minimum is just a lower value than nearby points

C. A global minimum is the absolute lowest point in the solution space, while a local minimum might not be the lowest point overall but is lower than all nearby points

D. A local minimum and global minimum are the same and represent the absolute lowest points in the solution space

Answer

The answer is *C*.

A local minimum is "*a point in the solution space where the objective function has a lower value than at all nearby points but may not necessarily be the absolute lowest point in the entire solution space.*" On the other hand, a global minimum is "*the absolute lowest point in the entire solution space, where the objective function has its smallest value.*"

Assessment

What is the key advantage of using SGD over basic gradient descent in the context of optimization, and how does it achieve this advantage?

Answer

The key advantage of using SGD over basic gradient descent is computational efficiency. SGD computes gradients and updates parameters using mini-batches of training data instead of using the entire dataset, making the process more efficient.

Understanding embeddings

At its core, an **embedding** is a mapping from a high-dimensional space to a lower-dimensional space that captures essential characteristics or features of data in a more compact form. This transformation not only reduces the dimensionality of the data but also helps NNs process and understand it more effectively.

These compact, meaningful representations of data play a pivotal role in various applications, from NLP to recommendation systems. In this section, we'll explore the concept of embeddings, their significance, and how they are employed to enhance the capabilities of NNs.

Word embeddings

Word embeddings are among the most renowned and widely used types of embeddings. They represent words as vectors in a continuous space, where each dimension of the vector corresponds to a semantic or syntactic feature of the word. This representation enables NNs to grasp meanings and relationships between words more intuitively.

Word embedding models generate word vectors by training on a large corpus of text data, learning to place similar words close to each other in the embedding space. Word embeddings have revolutionized NLP tasks, from **sentiment analysis (SA)** to machine translation, by providing models with a richer understanding of linguistic context.

Other embeddings include item (for example, images) and graph embeddings.

Training embeddings

Embeddings serve as the input layer in NNs, connecting raw data to the neural architecture. As the network learns during training, these embeddings may get adjusted to optimize model performance for the task at hand. Moreover, embeddings can be fine-tuned or kept static, depending on the problem requirements.

Training embeddings can take one of two approaches:

- **Pre-trained embeddings**: Pre-trained embeddings, such as Word2Vec or **Global Vectors (GloVe)**, are learned on massive datasets and can be used directly in NN architectures. They offer a valuable starting point for various tasks as they capture general relationships within the data.

- **Task-specific embeddings**: In some cases, embeddings may be trained specifically for a particular task or dataset. This custom approach tailors embeddings to a specific problem, potentially enhancing performance.

Assessment

In the context of embeddings and NNs, how do pre-trained embeddings differ from task-specific embeddings, and what is the potential advantage of using pre-trained embeddings?

Answer

Pre-trained embeddings, such as Word2Vec or GloVe, are learned on massive datasets and are used directly in NN architectures, capturing general relationships within the data. These embeddings provide a valuable starting point for various tasks due to their broad understanding of data relationships. In contrast, task-specific embeddings are trained specifically for a particular task or dataset, aiming to tailor the embeddings closely to that problem. The potential advantage of using pre-trained embeddings is that they offer a rich understanding of general data relationships, thus often speeding up training and possibly leading to better performance, especially when task-specific data is limited or lacks diversity.

Listing common network architectures

In the ever-evolving world of DL, network architectures serve as the blueprints for intelligence. Each architecture is a unique design, meticulously crafted to tackle specific challenges and excel in particular domains.

In this section, we'll embark on a journey through the diverse terrain of NN architectures, from CNNs, which conquer image analysis, to RNNs, which master sequential data, and from the creative minds behind **generative adversarial networks** (**GANs**) to the memory-enhancing capabilities of **long short-term memory** (**LSTM**) networks. Here, we'll list some common architectures and their applications.

Common networks

While explaining the distinctions between different network architectures is beyond the scope of this book, it is important to understand the basic differences between the most common networks. Here are some to keep in mind:

- **ANNs**: ANNs consist of interconnected nodes (neurons) organized in layers – an input layer, one or more hidden layers, and an output layer. Information flows forward through the network during inference, and backpropagation is used during training to adjust the weights to minimize the loss function.

- **RNNs**: RNNs are **sequence-to-sequence (seq2seq)** models, designed for processing sequential data such as text and time series data. They process sequences by maintaining a hidden state that carries information from the past. The hidden state is updated at each time step, allowing RNNs to capture dependencies over time. However, vanilla RNNs can suffer from the vanishing gradient problem.

- **LSTM networks**: LSTMs are a type of RNN that are designed to overcome the vanishing gradient problem. They use a more complex architecture with specialized gates (input, forget, output) to control the flow of information in and out of the cell state. LSTMs are well suited for modeling long-term dependencies in sequential data.

- **Gated recurrent units (GRUs)**: GRUs are another type of RNN architecture similar to LSTMs. They use gating mechanisms to control the flow of information within the network. GRUs are computationally more efficient than LSTMs and have been successful in various sequential data tasks.

- **CNNs**: CNNs are designed for processing grid-like data, such as images and videos. They use convolutional layers to automatically extract hierarchical features from the input. Convolutional filters slide over the input to detect patterns, and pooling layers reduce spatial dimensions. CNNs are widely used in image classification and computer vision tasks.

- **GANs**: GANs consist of two NNs – a generator and a discriminator – that are trained simultaneously. The generator tries to generate data that is indistinguishable from real data, while the discriminator aims to differentiate between real and generated data. This adversarial training process results in the generation of realistic data.

- **Graph convolutional networks (GCNs)**: GCNs are used for graph-structured data, such as social networks and molecular graphs. They generalize convolutional operations on graphs by aggregating information from neighboring nodes. GCNs can capture structural patterns and dependencies in graph data.

- **AEs**: AEs are a type of NN architecture that's used for **unsupervised learning** (UL) and dimensionality reduction. AEs find applications in tasks such as data denoising, anomaly detection, and FL. Variations of AEs, such as **convolutional AEs** (**CAEs**) and **variational AEs** (**VAEs**), have been developed to address specific types of data and learning objectives.

- **Transformers**: Transformers are a type of feed-forward neural network architecture that helped improve the shortcomings of sequence-to-sequence models like RNNs and LSTMs. The Vaswani et al. paper "*Attention is All You Need*" proposed transformers architecture with a mechanism called self-attention, which helped overcome the shortcomings in previously used sequence-to-sequence models like RNNs and LSTMs. These shortcomings include the vanishing gradient problem, and long-term memory loss due to its architectural design.

  The innerworkings of transformers are somewhat complex and thus beyond the scope of this chapter. However, it is important to note some of their architectural features and benefits:

  - **Encoder**: An encoder compresses the input data into a lower-dimensional representation, often referred to as a "latent space" or "encoding." This process captures the most important features and patterns in the data. The encoder uses self-attention and multi-head attention mechanisms. The encoder "encodes" both word vector embeddings and positioning information.

  - **Decoder**: A decoder reconstructs the input data from the lower-dimensional representation. The goal is to minimize reconstruction errors between the input and the output, encouraging the AE to learn a compact representation that retains important information.

  - **Encoder and decoder stacks**: Transformers often consist of stacked layers of encoders and decoders, allowing them to model complex seq2seq tasks effectively.

  - **Multi-head Attention (MHA)**: Transformers also use multiple-attention heads to learn multiple sets of weight matrices, producing more complex feature maps with multiple output channels. A multi-head attention mechanism simply allows the model to simultaneously learn multiple "types" of information from the same input. For example, an MHA mechanism might learn multiple pieces of information from the word "love", such as the context of the word, the part of speech the word represents, etc.

  - **Masked Multi-head Attention**: MHA may use masking techniques to improve the performance of a transformer. Masking is a method which "masks" words to improve its learning process. It effectively eliminates the model's dependency on "peaking" at future information, forcing it to identify additional patterns on less information.

  We will look at transformers and attention again later in the chapter.

Tools and packages

Python has firmly established itself as the *lingua franca* for researchers and practitioners alike. Its vast ecosystem of libraries, frameworks, and tools has made the development of NNs more accessible and efficient than ever before. Let's take a closer look at some of the most popular tools and packages that have become indispensable companions on the journey of building, training, and deploying NNs in Python.

TensorFlow, developed by Google, stands as one of the heavyweight champions of DL frameworks. Its flexibility, scalability, and extensive community support make it an ideal choice for both research and production environments. TensorFlow's high-level APIs such as Keras simplify the process of building and training NNs, while its lower-level operations allow for fine-grained control and optimization.

Keras, now an integral part of TensorFlow, has earned a reputation as the go-to library for building NNs with ease. Its high-level API abstracts many complexities, making it accessible to beginners and seasoned practitioners alike. With Keras, constructing intricate neural architectures becomes a matter of simple and expressive code.

PyTorch has gained immense popularity for its dynamic computation graph and intuitive interface. Developed by Meta's AI Research lab, it empowers researchers and developers to experiment with complex architectures and custom operations seamlessly. PyTorch's dynamic nature lends itself well to tasks involving variable-length sequences, RL, and generative models.

From DL frameworks such as TensorFlow and PyTorch to essential libraries for data manipulation and visualization, these tools provide a robust foundation upon which the future of AI is being built.

Assessment

Explain the main difference between the architectures of LSTM networks and GRU networks, and highlight specific use cases where each is beneficial.

Answer

LSTMs and GRUs are types of RNNs that are designed to handle sequential data, but they have different architectures:

- **LSTMs**: These have a more complex architecture with specialized gates – input, forget, and output gates. These gates control the flow of information in and out of the cell state. LSTMs were specifically designed to tackle the vanishing gradient problem, which can be a challenge in vanilla RNNs. The additional complexity of the LSTM allows it to model long-term dependencies in sequential data. LSTMs tend to be preferred for tasks where long-term dependencies of the data are critical, such as machine translation or speech recognition.

- **GRUs**: These are somewhat simpler in structure compared to LSTMs. Instead of having three gates like an LSTM, they use gating mechanisms to control the flow of information but combine the forget and input gates into a single "update" gate. GRUs are computationally more efficient than LSTMs due to their simpler structure. They have been successful in various sequential data tasks, especially when computational efficiency is crucial.

Introducing GenAI and LLMs

In the dynamic field of AI, language models stand as titans of NLU and generation. These models have not only revolutionized the way we interact with machines but have also sparked a renaissance in GenAI.

In this section, we'll delve into the world of LLMs, which are generative language models trained on massive text corpora (think in terms of most of the public data available on the internet) and can contain billions of parameters. We will focus on exploring LLMs: their architecture, training, and the transformative impact they have had on various applications, from text generation to chatbots, language translation, and even creative storytelling.

Unveiling language models

At their core, language models are **GenAI** models – these are AI models that generate texts, images, or other forms of media.

Specifically, language models are probabilistic models that learn the patterns, structure, and semantics of NL through NLP tasks. These models can predict the next word in a sentence, generate coherent paragraphs of text, and understand the meaning behind linguistic constructs – this is all thanks to their knowledge of language, which they gained through extensive training on large text corpora.

The impact of LLMs and GenAI reverberates across a multitude of domains:

- They have empowered chatbots to provide more natural and context-aware interactions

- They enable machines to translate languages, summarize texts, and generate human-like content

- They have become essential tools for creative writing, content generation, and even code completion, revolutionizing content creation and software development

The advent of transformer architecture (mentioned in the *Common networks* section) marked a turning point in the world of LLMs. As LLMs continue to evolve and grow in sophistication, they promise to bridge the gap between humans and machines in unprecedented ways. They have also shown an enormous potential to change the day-to-day reality of data scientists, who may spend less time building models from scratch and more time mastering the application and tuning of pre-trained models.

Furthermore, companies who wish to capitalize on the power of GenAI are eagerly seeking data scientists and AI engineers who have familiarity with this exciting new technology that has only become dominant in data science roles in the past few years. Thus, the journey into GenAI is far from over, and the stories, innovations, and applications it unfolds promise to be nothing short of extraordinary.

However, while LLMs and GenAI have opened doors to incredible possibilities, they have also raised concerns about ethics, bias, and misuse. The responsibility of ensuring that these powerful models are used for the greater good rests on the shoulders of researchers, developers, and society at large.

Transformers and self-attention

Transformers, which are neural network architectures using encoders and decoders, brought forth the concept of self-attention mechanisms, enabling models to capture long-range dependencies and contextual information efficiently (more on this in a second). They gained popularity after the release of "*Attention Is All You Need*" by Ashish Vaswani et al., published in 2017, and since then, they have become a cornerstone in NLP and various other ML tasks. Transformers are an improvement to the sequence-to-sequence models (seq2seq) like RNNs and LSTMs.

While the encoder is responsible for representing input data as vectors, the decoder is responsible for receiving and analyzing the output of the encoder and producing a sequential output. This is an appropriate architecture for NLP tasks such as text translation.

With transformers, the decoder can access additional hidden states, providing more "connections" or inputs for the decoder to decode.

Thus, the popularity of transformers almost seems to have arisen overnight, but they are in fact the result of years of DL architecture evolution. For example, while seq2seq models such as RNNs and LSTMs have been around since the 1990s (later enhanced with attention), transformers introduced the concept of "self-attention."

Let's take a look at the difference between the two:

- **Attention** is used in encoder-decoder transformer models, and calculates model weights using input queries and elements keys. These keys are then used to calculate weighted averages. The introduction of attention allowed the network to "remember" more information by connecting encoder outputs directly to decoder inputs. Think of the hidden state as a bottleneck like a toothpaste tube. You can only squeeze so much toothpaste (aka information) through the tube at a time. Attention was proposed as an extension to the encoder-decoder framework to connect information from one sequence (for example, input or encoder) to another (for example, output or decoder) directly, to produce predictions.

- **Self-attention** is like attention 2.0. Although similar, it has an important distinction. While attention allows transformers to access information from a different sequence, self-attention networks take this a step further by retaining an even larger context of information. This is achieved by connecting and learning information throughout the entire model architecture, creating multiple layers of weights of an input, which are then projected on the embeddings space. Instead of isolating the learning process within the encoder and decoder respectively, and then connecting them with attention, self-attention liberated AI from the seq2seq component entirely (although it can still be used to learn seq2seq tasks). In self-attention, the attention mechanism is used to encode information instead of seq2seq models such as RNNs by connecting multiple input variables throughout the network. Think of self-attention as a brain with multiple neuron connections throughout the entire brain as opposed to a system that only has a single highway between input and learned output. It would be much harder for us to learn information with such a limitation!

In short, self-attention allows each element in a sequence to consider all other elements when making predictions, capturing long-range dependencies efficiently.

Assessment

Describe the primary components and functionalities of transformers, and explain how they differ from traditional sequence-based models such as RNNs and LSTMs.

Answer

A transformer is a DL architecture that has become foundational in NLP and various other ML tasks. Its primary components and functionalities include the self-attention mechanism, which efficiently captures long-range dependencies; MHA, which allows the model to learn different types of relationships from the data concurrently; positional encoding, which gives the model a sense of order in the data; and encoders and decoders, which make the models great for complex seq2seq tasks.

Traditional sequence models such as RNNs and LSTMs process data in a sequential manner, with each step being dependent on the previous. In contrast, transformers can process all elements of a sequence in parallel thanks to the self-attention mechanism. Additionally, transformers, due to their self-attention mechanism, can capture long-range dependencies more effectively than RNNs or LSTMs, without worrying about issues such as the vanishing gradient problem.

Transfer Learning

After the introduction of transformers and self-attention networks, the realm of AI exploded with some of the most influential LLMs, including BERT, GPT, **Text-to-Text-Transfer Transformer** (T5), and their successors. These models became so powerful (in part because of their access to large corpora) that they gave rise to TL.

Transfer Learning (TL) is an AI technique in which a pre-trained model, initially trained on a large dataset for a specific task, is reused as a starting point for a different but related task. Therefore, instead of training a model from scratch, TL leverages knowledge and learned representations from the pre-trained model, allowing it to adapt more quickly to the new task.

TL is especially valuable when labeled data for the new task is limited, as it can significantly reduce the amount of data required for training. This approach has democratized AI development, allowing developers to leverage pre-trained models and adapt them to various applications.

GPT in action

As previously mentioned, GPT is one of the most popular pre-trained LLMs. Data scientists who used to build NLP tasks from scratch using Word2Vec embedding methods may now apply and fine-tune a GPT model, which already has a wealth of semantic language understanding. Thus, it's important to understand how to implement basic NLP tasks using GPT.

This section will provide some very basic implementations of text generation, **named entity recognition** (**NER**), and SA as a means to demonstrate the power of GPT using just a few lines of code. We encourage you to try more advanced examples in your LLM learning journey!

> **Note**
>
> In a real-world scenario, you would need to handle additional considerations such as model training, data preprocessing, and error handling. There are already entire texts dedicated to these topics. However, these examples are for illustrative purposes to provide a "crash course" on LLM implementation and aid you in LLM implementation conversations during interviews.

To get started, install the `transformers` library using `pip`:

```
pip install transformers
```

Now, let's have a look at three different examples.

Example 1 – Sentiment Analysis (SA)

SA is an NLP task involving extracting sentiment from a given text input. This is an example of analyzing the sentiment of a provided text:

```
from transformers import pipeline
nlp = pipeline("sentiment-analysis")
result = nlp("I love this movie!"[0]
print(f"label: {result['label'], with score: {round(result['score'],
4)}")
```

In this code, we do the following:

- Import the `pipeline` function from the `transformers` library
- Create an SA pipeline
- Pass text to the pipeline and index the result
- Print the sentiment prediction and its corresponding score using f-strings

Example 2 – Named Entity Recognition (NER)

NER is an NLP task involving extracting a named entity (for example, a person, place, and so on) from a given text input. This is an example of extracting a named entity from a provided text:

```
from transformers import pipeline

nlp = pipeline("ner")
result = nlp("Harrison Ford was in Star Wars.")
```

```
for entity in result:
    print(f"{entity['entity']}: {entity['word']}")
```

In this code, we do the following:

- Import the `pipeline` function from the `transformers` library
- Create an NER pipeline
- Pass text to the pipeline and index the result
- Print each recognized entity and its corresponding word in the text

Example 3 – Text generation

Text generation is an NLP task involving the generation of new text from a given input text. Here is an example of generating text provided some input text:

```
from transformers import GPT2LMHeadModel, GPT2Tokenizer

tokenizer = GPT2Tokenizer.from_pretrained("gpt2")
model = GPT2LMHeadModel.from_pretrained("gpt2")

input_text = "Once upon a time"
input_ids = tokenizer.encode(input_text, return_tensors='pt')

output = model.generate(input_ids, max_length=100, temperature=0.7,
do_sample=True)
output_text = tokenizer.decode(output[:, input_ids.shape[-1]:][0],
skip_special_tokens=True)

print(output_text)
```

In this code, we do the following:

- Import the necessary modules from the `transformers` library
- Load the GPT-2 model and GPT tokenizer
- Encode the text input into a machine-legible format
- Apply the model to generate text, specifying a maximum length and a temperature (which controls the randomness of the output)
- Decode the output from the model into human-readable text and print it

Summary

In this comprehensive exploration of DL, we embarked on a journey through the intricate landscapes of NNs, optimization algorithms, and fundamental concepts that underpin this transformative field. We began our voyage by deciphering NN fundamentals, understanding the building blocks of DL, and uncovering the power of activation functions, weight initialization, and embeddings. As we delved deeper, we navigated the seas of optimization, unraveling the intricacies of gradient descent, learning rates, and various optimization algorithms that guide the training of NNs. We also shed light on the vanishing and exploding gradient problems, which are crucial challenges to overcome in the pursuit of effective training.

Our odyssey continued with a tour of common network architectures, from CNNs mastering image analysis to RNNs and LSTMs excelling in sequential data tasks. We encountered the creative minds behind GANs, explored the power of transformers in NLU, and marveled at the capabilities of GCNs and GRUs. Transfer learning, auto encoders, embeddings, and the ethics of AI played pivotal roles in our journey, each adding a unique dimension to the ever-expanding universe of DL. We then explored the introduction of GenAI, particularly LLMs, and their evolution from seq2seq models with attention, to self-attention networks.

As we approach the shore of this chapter's conclusion, it's clear that DL is not merely a collection of techniques but a boundless realm of innovation and discovery. It empowers machines to comprehend and generate human-like intelligence, revolutionizing industries, research, and everyday life. As the tides of progress continue to surge, our voyage into the depths of DL is far from over, promising new horizons of understanding, creativity, and transformation in the ever-evolving world of AI.

In the next chapter, we will take our knowledge of how to build models to the next level by discussing model deployment.

12

Implementing Machine Learning Solutions with MLOps

Machine Learning Operations (**MLOps**) has emerged as a pivotal force in the data-driven age, enabling organizations to develop, deploy, and maintain machine learning models efficiently and effectively. It addresses key challenges related to speed, collaboration, governance, scalability, and cost, making it a discipline to be aware of for anyone navigating the modern landscape of artificial intelligence and machine learning.

In the following sections, we will break down the concept of MLOps, explore its core components, and provide insights into how it can elevate your machine learning initiatives. Whether you're an aspiring data scientist looking to see your models in action, an IT professional managing infrastructure, or a business leader shaping data-driven strategies, this chapter will equip you with the knowledge and tools you need to navigate the exciting and dynamic world of MLOps and have confidence in applying machine learning concepts to tackle data-driven challenges.

In this chapter, we will cover the following topics:

- Introducing MLOps
- Understanding data ingestion
- Learning the basics of data storage
- Reviewing model development
- Packaging for model deployment
- Deploying a model with containers
- Validating and monitoring the model
- Using **Azure Machine Learning** (**Azure ML**) for MLOps

Introducing MLOps

MLOps is an emerging discipline that blends the principles of DevOps and data science to streamline and enhance the machine learning life cycle. It encompasses a set of practices, principles, and tools designed to facilitate the entire journey of a machine learning model, from its inception to deployment, and beyond. In other words, MLOps is the bridge that connects the world of data science with the world of IT operations.

MLOps ensures that the promising machine learning models created by data scientists can be operationalized and maintained effectively in production environments. MLOps involves a holistic approach to managing machine learning workflows, covering aspects such as data acquisition, model development, testing, deployment, monitoring, and continuous improvement.

Why should you, as a reader, invest your time and energy in understanding and implementing MLOps? Here are some compelling reasons:

- **Efficiency and speed**: MLOps significantly improves the efficiency and speed of machine learning model development. It enables data scientists and machine learning/data engineers to iterate quickly and get models into production faster. This acceleration can be a game-changer for businesses aiming to stay competitive in rapidly changing markets.

- **Collaboration**: MLOps encourages close collaboration between data science and IT operations teams. This cross-functional cooperation ensures that the expertise of each group is leveraged effectively, leading to better outcomes and more successful projects.

- **Model governance**: In the era of data privacy regulations and industry standards, effective model governance is vital. MLOps provides the infrastructure needed to track and manage models, version data, and ensure compliance. This is particularly important for industries such as healthcare and finance, where regulatory requirements are stringent.

- **Scalability**: As machine learning models become more central to business processes, scalability is essential. MLOps helps organizations scale their machine learning workflows efficiently, whether it's deploying models across multiple regions, handling large volumes of data, or supporting more users and applications.

- **Cost reduction**: By automating repetitive tasks, optimizing resource utilization, and preventing costly errors, MLOps can lead to significant cost savings. It reduces the risk of downtime due to faulty models and minimizes the need for manual intervention in the deployment and monitoring processes.

- **Managing resources**: In addition to managing costs, there is a significant need to manage data from various processes (batch and streaming) across often complex data architectures, as well as managing code with version control.

If you ask someone what exactly MLOps entails, you'll get a million and two answers. This is because MLOps continues to be a very broad topic spanning roles, functionalities, and departments. While we can already assume that data scientists and data engineers are relevant to MLOps, you'll be surprised to learn that even IT and governance can be included in this massive process. However, if you're working for a smaller organization or start-up, you may discover that all of these roles are one and the same.

A model pipeline overview

A pivotal aspect of thriving in MLOps, a domain critical for modern data-driven organizations, is the mastery of crafting efficient and highly reproducible model pipelines. These pipelines aren't just a component of the workflow; they are also the backbone of a transformative approach in machine learning. By automating the intricate processes of building, training, and deploying machine learning models, these pipelines revolutionize the journey from a mere prototype to a robust production-ready solution. This automation not only dramatically accelerates the development cycle but also guarantees a consistent and error-free deployment, which is indispensable in today's fast-paced, data-centric world.

Developing model pipelines involves several essential steps and often relies on specific technologies to ensure reliability and consistency. You can see the data pipeline here:

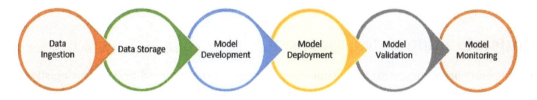

Figure 12.1: The data pipeline steps

This pipeline may look familiar because we've touched on most of these steps while learning about the ML workflow. However, there's so much more that goes on beyond the development and validation of the model.

Assessment

What is the significance of model pipelines in MLOps, and how do they contribute to the efficiency of machine learning workflows?

Answer

Model pipelines play a crucial role in MLOps, as they contribute to the efficiency of machine learning workflows in several ways:

- **Automation**: Model pipelines automate the complex processes of building, training, and deploying machine learning models. This automation speeds up the development cycle, making it possible to go from a prototype to a production-ready solution quickly. For example, an e-commerce company can use model pipelines to automate the recommendation engine's development and deployment, enhancing the user experience.

- **Consistency**: Model pipelines ensure consistency in model deployment. They guarantee that the same steps are followed every time, reducing the risk of errors and inconsistencies. In a healthcare setting, consistency is vital when deploying diagnostic models to ensure patient safety.

- **Reproducibility**: Model pipelines facilitate reproducibility by recording every step of the process. This is important in industries such as pharmaceuticals, where regulatory bodies require complete documentation of the model development process.

Now, in the following sections, we will take a look at each of these steps and the tools that are involved in each.

Understanding data ingestion

The responsibility of completing tasks within the early stages of the data pipeline (i.e., data ingestion and data storage) often falls under the responsibility of a machine learning/data engineer and not the data scientist. However, a data scientist should be able to understand what happens during these stages at a high level.

In the simplest terms, **data ingestion** involves developing automated processes to collect the data used for data science models automatically. Often, organizations/businesses already have processes in place to collect basic information about their activities, such as tracking website usage or customer purchase transactions. However, sometimes, to solve a particular organizational/business question, new data needs to be collected. The goal here is to automate the process to ensure that the data eventually used in a model is consistent, reliable, and free of bias to the best of the organization's ability.

Data ingestion usually occurs with ETL (**extract, transform, load**) or ELT (**extract, load, transform**) pipelines and typically involves batch and/or streaming processes. Going into depth about these two pipeline processes is outside the scope of this book; however, the important aspect to know is that these processes automatically collect data for an organization and output it, often in a structured format, ready for further processing or storage.

Here is a list of some of the technologies used during this step, each with their different strengths:

- **Apache Storm**: Apache Storm is a real-time stream processing system designed for handling high-throughput, low-latency processing of data streams. It's often used to process data as it arrives and can be integrated with other databases and message brokers.

- **Apache Beam**: Apache Beam is an open source, unified stream and batch processing model and SDK that allows developers to write data processing pipelines, running on multiple processing engines, including Apache Spark, Apache Flink, and Google Cloud Dataflow.

- **Hadoop**: Hadoop is an open source framework for the distributed storage and processing of large datasets, using a cluster of commodity hardware. It was developed by the Apache Software Foundation and has become a fundamental technology for handling big data. Hadoop is inspired by the Google File System and the MapReduce programming model, and it provides a scalable and fault-tolerant infrastructure to manage and process vast amounts of data.

- **Hive**: Hive is an open source data warehousing and SQL-like query language for Hadoop. It was originally developed by Facebook and is now maintained by the Apache Software Foundation. Hive provides a high-level interface to query and analyze data stored in Hadoop clusters, using a language similar to SQL called **Hive Query Language** (**HiveQL**). Hive allows users to create custom functions in Java, Python, or other languages to extend its functionality and perform complex operations. Furthermore, Hive integrates with various tools and frameworks in the Hadoop ecosystem, including HBase, Spark, and Pig.

- **Apache Spark**: Apache Spark is an open source big data processing framework that provides a unified and distributed computing engine for data processing. It's designed for speed and ease of use, making it suitable for large-scale data preprocessing and transformation tasks. It uses an in-memory processing model to process data in parallel across a cluster of computers and employs a **directed acyclic graph** (**DAG**) execution model to optimize data processing workflows. Spark's core data structure is the **Resilient Distributed Dataset** (**RDD**), which is fault-tolerant and allows for parallel processing. In Python, you can use Spark with the PySpark API.

- **Dask**: Dask is a versatile and powerful data processing framework, but it's unique in the sense that it can handle both batch and stream data processing, making it a great choice for a wide range of data processing tasks. It is an open source parallel computing library in Python that can handle larger-than-memory datasets. It is designed for parallel computing and distributed computing tasks. Similar to Spark, Dask breaks down complex tasks into smaller, manageable operations that can be parallelized. It leverages parallel computing frameworks such as threading, multiprocessing, and distributed computing to process data in a distributed and scalable manner.

Although Spark (`https://spark.apache.org/`) and Dask (`https://www.dask.org/`) are beyond the scope of this chapter, it's worth looking into the documentation of both frameworks to understand the program syntax. If you're pretty comfortable with Pandas, you'll be up and running in Spark and Dask in no time!

Now that you have the data ingested, let's discuss how you might want to organize and store it.

Learning the basics of data storage

As stated earlier, the data storage step in the model pipeline process tends to be a function of machine learning/data engineers. However, it is beneficial for a data scientist to have a basic understanding of this step.

Data storage is simply about housing the data that we gather from different sources. There are a variety of approaches to this, depending on the data's requirements (e.g., the structure, schema, size, ingestion type, privacy, etc.).

The following are some examples of data storage options within MLOps:

- **Binary Large Object (BLOB) storage**: BLOB storage is a type of data storage that is designed to store and manage large binary data, such as images, videos, documents, and other types of files. BLOBs can be of varying sizes, from small to very large, and they are typically unstructured data, meaning they lack a specific schema or organization. In modern data architectures, the cloud services offered by Azure Blob Storage, Amazon **S3** (**Simple Storage Service**), and Google Cloud Storage are used to store and manage BLOB data. These services are highly scalable, durable, and optimized for web and cloud-based applications.

- **Traditional databases**: As you've already learned, traditional, structured databases are **relational database management systems** (**RDBMSs**) that use a structured and tabular format to store and manage data. SQL is both a language and a set of conventions used to define, query, and manipulate data within these databases. SQL databases are widely used in various applications and systems to manage structured data efficiently.

- **Graph databases**: Graph databases are a category of NoSQL databases designed to store and manage data in the form of graphs. In graph databases, data is structured as nodes (vertices) and edges (relationships), allowing for the representation and storage of complex and highly connected data. These databases are particularly well-suited for data models where relationships between entities are as important as the entities themselves. In a graph database, data is organized into a graph, which consists of nodes and edges. Nodes represent entities (such as people, products, or locations), and edges represent the relationships or connections between these entities. They often come with their own query languages, such as Cypher (used in Neo4j), Gremlin (used in Apache TinkerPop), and SPARQL (used in RDF databases).

So far, data has been collected, organized, and stored. It's now ready for model development, where you can flex your data scientist muscles to develop an awesome model.

Reviewing model development

Model development includes discovering relationships between data and features and better understanding the context of the business question being solved. This may also be a good time to understand KPIs and success measures, as well as the overall structure of the business problem. Performing descriptive statistical analysis and creating data visualizations are also ideal activities at this stage of the pipeline.

As you learned in previous chapters, you can perform data analysis and model development in Python, as well as R. Python offers a number of useful packages that we've already discussed, including Keras, TensorFlow, and PyTorch. There are also "auto-ML" frameworks where models can be developed and run in the cloud, including Google AutoML, Azure ML Studio, Amazon SageMaker, IBM Watson, Databricks AutoML, H20, and Hugging Face.

We will skip over the details of ML development, since we already discussed them at length in the chapter on machine learning in *Chapter 10*. However, it is worth noting an important concept that we did not discuss – experiments.

Experiments are systematic and structured trials or tests that you conduct during the model training and evaluation process. In *Chapter 10*, we talked about model tuning, where you adjust different model hyperparameters to find the optimal combination. Experiments allow you to do this. For example, you might run different experiments to test how the number of random forests impacts your results. You have already been informally exposed to experiments during the localized model-tuning process on your machine.

However, when tuning models in the cloud, you can systematically track the performance of each experiment with specific model architectures, features, and sets of hyperparameters. This process also involves tracking model performance metrics.

Open source options for tools to run experiments and track the results include MLflow, **Weights & Biases** (**W&B**), **Data Version Control** (**DVC**), and Guild AI. The advantage of automating model hyperparameter tuning through code is that it can be integrated into your model-training MLOps pipeline. Consequently, you can easily rerun these experiments as needed in the future for retraining purposes. Additionally, this approach documents the process of selecting the best model.

Choosing a model that meets or outperforms a given threshold is useful to determine the best-fit model. Once the best model is chosen, stress-testing it (for example, by giving it specific test data that it may encounter in the real world) and automating unit testing are also typically part of the model development process. Now, let's turn our attention to model deployment and how to package the model for deployment.

Packaging for model deployment

Once you're happy with the model that you've chosen in the model development process, it is time for the model deployment process! However, before deploying the model, it is important that it's properly packaged for production. There are a number of approaches to packaging an ML software program, but we will review the version that you are more equipped to learn – Python pip packages.

pip is the standard package manager for Python, and it is used to install, upgrade, and manage Python libraries and dependencies. A Python **pip package** refers to a software package that can be easily installed and managed using the pip package manager.

Most Python packages are hosted on the **Python Package Index** (**PyPI**), which is a repository of Python packages that can be easily accessed and installed using pip. These packages are designed to be libraries or reusable modules that can be imported and used in other Python scripts or projects. The main functionality of the package is organized in Python modules and can be accessed by importing them, but there is no specific "main" script like you might find in a standalone application.

A pip package typically consists of one or more Python modules, scripts, or other resources that provide specific functionality. These packages are created and distributed to facilitate the reuse of code and to allow developers to easily integrate them into their projects.

These pip packages can take many forms depending on the project; however, when discussing packaging for deployment, it's important to consider any requirements for your code and the environment it runs in for your model to run correctly. We will discuss them in the upcoming subsections.

Identifying requirements

One important point about packaging your model for deployment includes identifying the requirements to run the model. For example, does your model script require the Python packages of NumPy, Pandas, or scikit-learn to run correctly? If so, what versions of those packages? What version of Python is required?

When building a pip package, you can define some of these requirements in the `Requirements.txt` files. This is a configuration text/flat file that specifies all the versions of each package you want to use. Then, when a teammate runs your code, the code references the correct packages and their versions.

Now that you've defined the requirements for your model, we should start to think about the environment that you run your model in.

Virtual environments

As we journey deeper into the world of MLOps, particularly in the context of deploying ML models, one significant aspect stands out – the creation and management of your environment using code. Often, when using the cloud, you can use code to define what type of computing resources you want your model to run on. For example, you can write in code that you want to deploy your model on a computer running the exact version of Python you identified in your requirements section for your model. This practice, often referred to as **Infrastructure as Code (IaC)**, is a key strategy that allows data scientists, particularly those venturing into MLOps, to handle environments where ML models run efficiently and reliably.

Understanding the benefits of defining environments through code is crucial. Firstly, it ensures consistency. By codifying the environment, you ensure that your model runs in a controlled and predictable setting, reducing "it works on my machine" syndrome. This consistency is vital when moving models from development to production, where differing environments can lead to unexpected behaviors in models.

Moreover, using code to define environments enhances collaboration and version control. Teams can share, review, and update environment configurations just as they would with source code, making collaborative work more streamlined. This approach also integrates smoothly with version control systems such as Git, allowing you to track changes and maintain a history of an environment, just like you would with your models and data pipelines.

Tools and approaches for environment management

Several tools and approaches can be employed to manage environments in code.

For containerization, Docker (which we will talk about shortly) is a popular choice, allowing you to package your application and its dependencies in a container that can run on any system. This encapsulation ensures that your model has all the necessary libraries and settings, irrespective of where it's deployed.

To orchestrate these containers, particularly in more complex deployments, tools such as Kubernetes can be invaluable. "Orchestrate" in this context refers to the coordinated management and control of multiple containers, ensuring they work together seamlessly. Kubernetes helps manage and scale your containers across multiple machines, handling tasks such as load balancing and fault tolerance. It's particularly useful when deploying models at scale.

On the infrastructure side, tools such as Terraform or AWS CloudFormation allow you to define cloud resources as code. This means you can create, modify, and manage the cloud infrastructure that supports your models in a repeatable and automated manner. By using these tools, you can easily replicate your production environment for testing, ensuring that your models behave as expected when deployed.

As you progress from defining model requirements to deploying them, integrating environment management into your workflow is a logical next step. By treating your environment as part of your code base, you align it with the core principles of MLOps – reproducibility, scalability, and maintainability. This approach not only simplifies the deployment process but also paves the way for more robust and reliable ML systems. Container software tools such as Docker and Kubernetes are popular for managing the model environment. Let's discuss containers more as we discuss model deployments.

Deploying a model with containers

In the world of MLOps, containers have become a cornerstone for deploying ML models, offering a lightweight, consistent, and scalable solution for running applications, including ML models, across various environments. Containers encapsulate an application, its dependencies, and runtime into a single package, ensuring that the model behaves the same way regardless of where it is deployed.

This is particularly important in MLOps, where models need to perform consistently across development, testing, and production environments. Once the model is containerized, it can be deployed to a variety of platforms. Cloud services such as **Azure Kubernetes Service (AKS)** or **Amazon Elastic Kubernetes Service (EKS)** can be used to manage and scale containers.

Containers address several key challenges in MLOps. First, they solve the "it works on my machine" problem by providing an isolated environment that is consistent across all stages of the deployment pipeline. Second, they facilitate scalability and load balancing, which are crucial for handling varying demands in production. Lastly, containers enhance collaboration among team members by ensuring that everyone works in a consistent environment, reducing conflicts and speeding up the development process.

Now that you know more about some of the benefits of containers, let's turn our focus to a very popular containerization tool – Docker.

Using Docker

Docker is a very popular tool for creating and managing containers. It allows you to define your environment and dependencies in a Dockerfile, which can then be used to build a container image. Here's a basic example of a Dockerfile for an ML application:

```
# Use an official Python runtime as a parent image
FROM python:3.8-slim
# Set the working directory in the container
WORKDIR /usr/src/app
# Copy the current directory contents into the container
COPY . .
# Install any needed packages specified in requirements.txt
RUN pip install --no-cache-dir -r requirements.txt
# Make port 80 available to the world outside this container
```

```
EXPOSE 80
# Define environment variable
ENV NAME World
# Run app.py when the container launches
CMD ["python", "app.py"]
```

In this Dockerfile, we define a Python environment, set up the necessary files, install dependencies, and specify how the application should run. We then tell Docker to run the Python program called app.py once the container launches.

Let's assume that app.py contains code you have written to take in input from the outside world and process it with your trained model to return a result. Once the container is up and running, this makes your model available to start churning out predictions. However, at this point, the model is not running yet because all we have done is give Docker a list of instructions. We still have to build and run the container.

Once the Dockerfile is defined, you can build and run a container using Docker commands. Here's how you do it:

```
# Build the Docker image
docker build -t my-model .
# Run the Docker container
docker run -p 4000:80 my-model
```

This builds a Docker image named my-model and runs it, mapping port 80 of the container to port 4000 of the host machine. Your model should now be up and running and ready to accept input.

To summarize this section, in an MLOps pipeline, containers are typically used in the training and deployment phase. After a model is developed and trained, it is packaged into a container. This container can then be deployed to various environments (such as testing, staging, and production) without any changes, ensuring consistency across the pipeline. For more complex applications, especially those requiring scalability and high availability, you might use Kubernetes in your MLOps deployment process to manage the automation and deployment of containers across a cluster of machines, such as the Docker container we just discussed.

Assessment

What is Docker, and how does it contribute to the containerization process in MLOps? Also, discuss why containers are so important to the MLOps process.

Answer

Docker is a popular containerization tool used extensively in MLOps, simplifying the process of creating and managing containers.

In the realm of MLOps, containers play a pivotal role, and they have been crucial for deploying ML models effectively. These lightweight, portable, and self-contained units package not just the model but also its dependencies and runtime environment. This encapsulation ensures that an ML model behaves consistently, regardless of the environment which it's deployed in. Additionally, collaboration is at the heart of MLOps, and containers facilitate it by offering a standardized environment for all team members. This harmonizes efforts, reduces conflicts, and accelerates the development cycle.

Validating and monitoring the model

After you've successfully trained and deployed your ML model, the journey doesn't end there. Model validation and monitoring are the important next steps in your MLOps process. We will briefly discuss validating your deployed model and then focus on monitoring it long-term.

Validating the model deployment

Once your model is deployed, you will want to validate that it works as expected. This is a relatively short and straightforward process. The general steps involve connecting to your deployed model, submitting some data (preferably data unseen by the model during the training process), collecting the model predictions, and scoring them.

This will allow you to confirm a couple of things. First, you know that your deployment worked, and your model is returning results. Secondly, if you submit unseen data to the model and score it, this will give you another assessment of the model's performance. You don't want to be surprised by it. Thus, it is a good idea to check that you're getting the results that you expected.

Assuming that your model is deployed with Docker, here is a sample of how you might validate your deployed model (we are only providing pseudocode because the details of your code will depend a lot on the context, such as how your model is deployed and the types of input it will expect):

```python
import requests

# Prepare unseen data (ensure it has the same features as the training
data)
unseen_data = ...

# Get the IP address of the container
ip_address = ...

# Make predictions on the unseen data
response = requests.post(f'http://{ip_address}:port/predict',
json={'data': 'unseen_data'})

# Evaluate model performance (e.g., calculate accuracy or other
metrics)
...
```

This code highlights gathering data unseen by the model during its training, finding out the IP address of your deployed container, and then submitting the data. Finally, the response value is evaluated for its performance. If your code cannot complete the steps of producing a prediction or the prediction values aren't as expected, you know you have an issue with your model or model deployment.

Once you've validated that your deployed model performs as expected, then you will need to think about monitoring it.

Model monitoring

Model monitoring is a crucial aspect of the ML life cycle, involving tracking, analyzing, and maintaining your models to ensure they continue to perform well in production. In Azure ML, you can implement model monitoring as part of your broader MLOps pipeline.

Imagine you've built a model that predicts customer preferences for an e-commerce platform. Initially, everything seems perfect; your model is making accurate recommendations, and everyone is happy. But what happens when, over time, the data changes, user behavior evolves, or unexpected errors occur? Without proper monitoring and logging in place, you'd be operating blindly, unaware of these critical shifts.

Logging is the practice of recording events and activities related to your model's operations. Think of it as a journal that documents every interaction and decision made by your model. Why is this important? Logs serve as a historical record, helping you trace back and understand what happened when issues arise. They are your detective tool for troubleshooting and debugging. Using these logs, you can also monitor your deployed model.

Monitoring your deployed model provides real-time awareness of its performance. It's like having a dashboard that tells you how well your model is doing at any given moment. You can track metrics such as accuracy, response times, and resource utilization. When something goes awry, such as a sudden drop in accuracy or increased response times, monitoring alerts you immediately, allowing you to take corrective actions swiftly. In addition to monitoring your model's performance metrics, you also want to monitor the input data for data drift.

Data drift occurs when the statistical properties of the input data that your model receives change over time. These changes can be subtle or significant, and they can impact your model's performance. You can detect data drift by establishing a baseline for your model's performance on your initial training data. This baseline serves as a reference point for future evaluations. Then, regularly compare incoming data with the data your model was trained on.

Statistical tests and techniques such as the population stability index, Jensen–Shannon divergence, or simple feature statistics can help detect changes in data distribution. Based on the drift detection or performance metrics, you may need to retrain your model. However, by following the principles of MLOps, where most of the model-building process has been coded and automated, the retraining process should be fairly easy.

We have now covered at a high level what model monitoring looks like. However, now that you have a model in production, you might also want to think about ML/AI governance as another aspect of model monitoring.

Thinking about governance

Congratulations! You now understand the high-level steps associated with deploying an ML model. In many cases, this is where the "data science workflow" concludes. This is not to say that data scientists aren't involved with down-the-line activity from this model, such as preparing memos or educational material on the model design and performance, but rather that the bulk of the work associated with most data science jobs has been covered. However, when it comes to monitoring your model, you might want to consider a broader perspective that includes the system and how it's governed. Technically, a data scientist can implement and deploy a model, but there may be questions or concerns about how it's used later on. This is where governance becomes important.

A data scientist who expresses their knowledge and commitment to ML/AI governance will certainly stand out from the crowd. Employers want data scientists who go the extra mile by considering business context, needs, and concerns.

ML/AI governance refers to the set of policies, processes, and practices established to oversee and manage machine learning or artificial intelligence systems and their operations, within an organization or a broader ecosystem. It involves defining rules, regulations, and ethical principles that guide the development, deployment, and use of AI technologies to ensure responsible, fair, and secure outcomes.

Key aspects of AI governance include the following:

- **Ethical guidelines**: Establishing guidelines that prioritize fairness, transparency, and accountability in ML/AI systems. This includes addressing potential biases, discrimination, and ethical concerns in AI applications.

- **Data privacy and security**: Ensuring that ML/AI systems handle data in compliance with privacy regulations and industry standards. Protecting sensitive information and mitigating data breaches is a fundamental component.

- **Compliance and regulations**: Adhering to legal requirements and regulations related to ML/AI, data, and cybersecurity. Compliance with industry-specific standards and international laws is essential.

- **Accountability**: Defining roles and responsibilities for ML/AI system developers, operators, and users. Accountability ensures that any issues or challenges can be appropriately addressed.

- **Transparency and explainability**: Demanding transparency in ML/AI decision-making processes and making AI model predictions explainable to build trust and facilitate human understanding.

- **Monitoring and auditing**: Implementing mechanisms to continuously monitor ML/AI system performance, assess its impact, and conduct regular audits to ensure adherence to governance principles.

- **Risk management**: Identifying and mitigating potential risks associated with ML/AI, including security vulnerabilities, ethical concerns, and compliance gaps.

ML/AI governance is an evolving field, as technologies advance and organizations strive to address the challenges and opportunities presented by this new technology. Admittedly, some industries have more mature governance policies than others (e.g., healthcare, insurance, and finance). However, as you can see, it plays a crucial role in balancing innovation with responsibility and ensuring that ML/AI benefits society as a whole, while minimizing potential harm.

Using Azure ML for MLOps

There are many different platforms for orchestrating your MLOps. Here, we will just focus on one tool, Azure ML. As a comprehensive cloud-based platform, Azure ML can play a significant role in various stages of the MLOps pipeline, fitting seamlessly into your existing framework of data ingestion, storage, development, deployment, validation, and monitoring. Here's how Azure ML integrates with each of these stages:

1. **Data ingestion**: Azure ML supports various data sources, allowing for flexible data ingestion. It can connect to Azure Data Lake, Azure Blob Storage, and other external sources. This flexibility ensures that data ingestion, a critical first step in the MLOps pipeline, is streamlined and efficient.

2. **Data storage**: With Azure ML, data storage is integrated with Azure's cloud storage solutions. It allows for the secure and scalable storage of large datasets, essential for ML workflows. This integration facilitates easy access and management of data within the MLOps pipeline.

3. **Model development**: Azure ML shines in model development with its wide range of tools and capabilities, including Jupyter notebooks, **automated machine learning** (**AutoML**), and support for various ML frameworks. It provides an environment where data scientists can experiment, develop, and iterate models efficiently.

4. **Model deployment (using Azure ML as an example)**: Azure ML excels in model deployment, offering tools for easy deployment of models as web services in the cloud or on the edge. It simplifies the process of deploying models into production, handling the complexities of scalability, load balancing, and security. By using Azure ML, you can demonstrate how models can be packaged, versioned, and deployed, maintaining consistency across different environments.

5. **Model validation**: Azure ML supports model validation processes through its robust testing and evaluation tools. It allows you to create various validation scenarios, track performance metrics, and compare different model versions. This ensures that only well-performing models are deployed.

6. **Model Monitoring**: Post-deployment, Azure ML offers powerful monitoring capabilities. It tracks the performance of models in production, detects data drift, and provides insights into model behavior. This monitoring is crucial for maintaining the accuracy and reliability of models over time.

In summary, Azure ML is not just a tool for model deployment; it's an end-to-end platform that supports the entire MLOps life cycle. Its integration at each stage of your MLOps pipeline can enhance the efficiency, scalability, and effectiveness of ML workflows.

Summary

In this high-level introduction to MLOps, a crucial discipline in the AI and data science landscape, we delved into its key aspects. We began by understanding the significance of MLOps, its role in bridging the gap between model development and production deployment, and the impact of a well-structured MLOps pipeline on business outcomes.

The chapter covered the MLOps journey, emphasizing the importance of reproducibility, collaboration, and automation in the ML workflow. We explored developing model pipelines, technologies such as Docker and Databricks, and model versioning. Additionally, we discussed the cloud-native tools and services available to manage ML experiments and monitor model performance. Finally, we examined governance and compliance practices in AI, ensuring ethical and regulatory alignment.

This chapter serves as a roadmap for implementing MLOps best practices, enabling organizations to develop, deploy, and manage ML solutions efficiently and responsibly in today's data-driven world.

Now, we will conclude the technical content of the book to help you prepare for your technical interview. The remainder of the book will focus on other non-technical aspects of the interview. In the next chapter, we will focus on interview preparation and what types of questions you might expect from a recruiter or hiring manager.

Part 4:
Getting the Job

The final part of this book aims to provide tips and insight for data science interviews. You will learn how to best prepare for them, and how to effectively negotiate your salary and benefits. At the end of this section, you will have valuable knowledge on how to succeed in your job search and how to optimize your outcomes.

This part includes the following chapters:

- *Chapter 13, Mastering the Interview Rounds*
- *Chapter 14, Negotiating Compensation*

13

Mastering the Interview Rounds

So, at this point, you've explored the data science landscape, the fundamentals of programming in Python, the puzzling world of SQL queries, the wonder of data visualization and storytelling, and the productive advantages of leveraging the command line and Git. You then jumped head-first into the concepts of statistics, pre-modeling tasks, **machine learning**, **neural networks**, and model deployment. You've basically undergone a crash course in data science 101, covering about 99% of what you'll encounter in data science interviews. Now what?

You're probably wondering what to expect if you've never interviewed for a data science role. Well, here's the thing: the interview process for a data science position in one organization can be very different from another. However, there are some commonalities that we will review. Additionally, the content covered in the earlier chapters of this book should put you in a great position to do well in your interview too.

So, in this chapter, we will review the experiences that you can bet on having in the data science interviewing world, the basic anatomy of the interview process, and what to expect from each stage. This includes the following:

- Mastering early interactions with the recruiter
- Mastering the different interview stages
- Mastering the hiring manager stage

Mastering early interactions with the recruiter

In *Chapter 2*, we shared some tips on how to optimize your data science job search. In this section, we'll discuss what to expect when you receive your first recruiter inquiry. Hooray!

Recruiter screenings are often the first stage in most corporate interviews. It involves someone from a company's recruiting team reaching out to you for an initial conversation regarding the role. If you receive a call, a message, or an email from a recruiter, you should pat yourself on the back because the following statements are now officially true:

- *You're qualified*: Recruiters don't call applicants who aren't qualified. So, you can celebrate a) having the necessary skills for the role and b) writing an engaging resume and/or cover letter that effectively speaks to said skills.

- *You're in the top 2% of applicants*: Some research suggests that corporate roles receive an average of 250 applications (at least before they stop collecting applications!). Of those, only 4 to 6 receive callbacks [1].

But despite your fortune, the odds are still against you. You've come a long way, but not far enough. You're now tasked with besting 3-5 others who are just as (if not more) qualified as you are. This means proving to the hiring team that you are the best fit, not just someone who could fit (a message you're trying to convey throughout the entire interview process).

The introductory call/message/email will likely entail the recruiter asking you for your availability, so be sure to provide at least three time blocks. The sooner you can bypass the recruiter screening, the better.

> **Note**
>
> It's important to schedule your interview ASAP. Although you were called into the interview process, you may be one of the last candidates to get invited. This means candidates further along in the process have an advantage. If they impress the interviewers enough, it's not unusual for the recruiter or hiring manager to cut the process short.

Although the recruiter screening is relatively simple to bypass, it's also easy to underestimate. To maximize your performance in this interview, prepare to answer the following questions:

- *What do you know about our company?*: Be sure to research the company before the recruiter screening. This should go beyond the basic business model of the business, including strategic bets, recent news developments, and challenges the company or industry may be facing. For public companies, you can often find this information in the company's latest annual report and press releases; for start-ups and private companies, however, it can be harder to find this information.

- *Why are you interested in the role?*: Provide a brief description of what attracts you most to the opportunity. It's good if you can fit in a factoid about the company that attracted you, preferably something beyond what you can discover from a basic internet search on the company. This shows strong compatibility with the role and your due diligence in company research.

  Also, feel free to tie something in the job description to your personal interests or experiences. For example, if you're applying for a consumer-facing role and are interested in ensuring results from ML models are unbiased, connecting your interest to how it could impact everyday consumers would be a great way to show your interest in the role.

- *Walk me through your background and what you're working on today*: Many make the mistake of assuming this question is to learn about your background and interests. While this is true, it's much deeper than that. The recruiter is specifically seeking to find out how your experience meets the job description. Spend this time explicitly mentioning jobs, projects, and achievements relating directly to the role for which you're interviewing.

In addition to these questions, the recruiter will likely ask you logistical and basic qualification questions about work authorization, willingness to travel, preferred location, and/or the work format (remote versus hybrid versus on-site), and so on.

Note

The key to maximizing your odds of landing a role depends on maximizing your job opportunities. This often involves removing filters from your job search that otherwise limit your options. One major filter to consider removing is the work format. While there are still many remote roles, they receive more applicants than on-site roles. If you're seeking your first data science job, it might be worth sacrificing your preferred work format in order to optimize your chances. Going into the office is also a great way to establish stronger relationships for career growth and boost your opportunity options.

In general, recruiter screenings are no longer than 20 minutes long. This is because they're simply validating that what you've shared in your application is legitimate and that you're capable of (and still interested in) pursuing the role. It is also an opportunity for the recruiter to report back on your key qualifications to the hiring manager. In short, the recruiting stage is about maintaining and enhancing expectations that the recruiter already has regarding your candidacy.

Once the screening concludes, the recruiter will share their notes with the hiring manager. This candidate summary will dictate whether you're worth moving to the next stage. This is why it's absolutely important to provide STAR examples of your experience.

The **Situation, Task, Action, and Results (STAR) method** is an interview framework used to structure behavioral interview questions aimed at investigating a candidate's work style and ethics in the form of work experience, critical thinking, outlook, attitude, accomplishments, and technical rigor (to name a few).

Although major companies such as Amazon, Walmart, and McKinsey recommend its candidates' responses follow the STAR method, it's a generally strong framework to leverage, regardless of the company or even the role. Let's break it down:

- *Situation*: Begin by describing the context or situation you were in. This sets the stage for your story. Provide enough detail to help the interviewer understand the scenario you faced.

- *Task*: Explain a specific task or challenge you needed to address in that situation. What were you required to accomplish or solve?

- *Action*: Detail the actions you took to tackle the task or resolve the situation. Focus on what you did, emphasizing your individual contribution. Describe the steps you took, the skills you used, and the decisions you made.

- *Result*: Conclude by outlining the outcomes or results of your actions. What happened as a result of your efforts? It's crucial to highlight the positive impact you had, whether it was achieving a goal, improving a process, or solving a problem.

Here is an example of how you might use the STAR framework to answer the question, *Tell me of a time when you needed to handle a pressure situation*:

- *Situation*: "Once, while attending an important meeting, I noticed a mistake made by my manager in a solution our team was developing."

- *Task*: "If the error had gone unaddressed, there would have been significant financial ramifications and potentially a loss of customer trust."

- *Action*: "I requested to speak with my manager one-on-one after the meeting to address the situation. I politely raised my concern in a private setting to avoid any issues of embarrassment."

- *Result*: "After reviewing the issue, my manager agreed with my assessment that there was an error, and we quickly issued a correction, avoiding any financial losses for the organization."

As you can see, using the STAR method helps structure your responses and provides a clear framework for showcasing your skills and experiences. It allows interviewers, including recruiters, managers, and panelists, to understand how you approach challenges and assess your ability to handle different situations.

Mastering the different interview stages

After successfully passing the recruiter screening, where your qualifications and initial fit for the role are evaluated, the journey intensifies as you enter the more challenging stages of the interview process. This next phase is not just a continuation but a significant escalation in the evaluation of your skills.

In addition to the hiring manager stage, where your fit within the company's culture and soft skills are thoroughly examined, you'll also face the technical interview. During the technical interview, your specific skills and competencies relevant to the role are rigorously assessed. Furthermore, you may encounter panel interviews, where multiple key stakeholders, including potential future colleagues, assess your ability to contribute to the team dynamically.

This comprehensive approach ensures a holistic evaluation of your technical prowess, behavioral traits, and compatibility with the company's ethos.

The hiring manager stage

The **hiring manager interview** is one of the most (if not *the* most) important stages of the interview process. It likely consists of both behavioral and technical inquiries to assess your fit to the role and team. It is also an opportunity to express your interest in the role and why you are such a great fit. In short, your goal should be to highlight why you're the best candidate and to address any concerns or assumed gaps in your candidacy.

Thus, reaching the stage of interviewing with a hiring manager typically indicates several positive assumptions about your stance in the interview process:

- *Fit for the company culture*: If you reach this stage, you likely align well with the company's values, mission, and work culture.

- *Technical competency*: You have likely demonstrated technical competencies or skills relevant to the role. The hiring manager may focus on deeper discussions about these skills during the interview.

- *Strong communication and soft skills*: Reaching this stage suggests you possess strong communication and interpersonal skills. Hiring managers then often gauge how well you can articulate your thoughts, engage in discussions, and handle various scenarios.

- *Your interest*: Making it to the hiring manager stage indicates your genuine interest in the position and company. You've likely shown commitment and enthusiasm throughout the earlier stages.

During an interview with a hiring manager, you can expect the following:

- *Deeper technical or role-specific questions*: The discussion may involve more detailed questions related to the specific job responsibilities and technical skills required.

- *Assessment of fit*: The hiring manager might delve into how you fit within the team dynamics and the broader company culture.

- *Behavioral and situational questions*: Expect questions about past experiences and how you handled certain situations. Again, the STAR method might be used to structure your responses.

- *Discussion on career goals and aspirations*: The hiring manager might inquire about your long-term career goals and how they align with the role and the company.

- *Final assessment*: Sometimes, this stage serves as a final evaluation before a hiring decision (if there is no requirement to meet other interviewers). The hiring manager will assess whether you are the best fit for the role and the team, typically with a take-home assignment, technical assessment, or presentation.

When you reach the hiring manager stage, you should prepare to showcase your technical skills, personality, cultural fit, and enthusiasm for the position and the company. However, organizations looking to hire you are likely to be interested in not only your cultural fit but also your technical acumen. Therefore, you should be prepared for the technical interview. The following section talks about what you might expect in a technical interview.

The technical interview

Encountering the **technical interview stage** in a data science hiring process indicates that you have demonstrated foundational skills and qualifications, advancing to a phase that specifically evaluates technical expertise and problem-solving abilities.

At this point, assumptions about your candidacy include the following:

- *Technical competence*: If you've reached the technical interview stage, it's likely that you possess a solid foundation in data science concepts, statistics, programming languages such as Python or R, ML algorithms, and data manipulation techniques

- *Problem-solving skills*: You've likely showcased your ability to solve complex data-related problems and analyze datasets effectively in earlier stages, leading to this phase

- *Understanding of algorithms and models*: You've shown a thorough understanding of various ML algorithms, statistical models, and their applications in real-world scenarios

- *Programming Proficiency*: Proficiency in coding and data manipulation using libraries such as Pandas, NumPy, scikit-learn, or TensorFlow is assumed at this stage

To excel in a technical data science interview, consider the following tips and best practices:

- *Review core concepts*: Ensure a strong grasp of fundamental data science concepts, including statistical methods, ML algorithms, data preprocessing, model evaluation, and feature engineering.

- *Practice coding*: Practice coding in Python or R extensively. Be able to solve data science-related problems on platforms such as LeetCode, HackerRank, or Kaggle to improve coding skills and algorithmic understanding.

- *Understand model implementation*: Be prepared to discuss and implement ML models, their advantages, limitations, and scenarios where they're most suitable.

- *Showcase projects*: Highlight personal or professional projects that demonstrate data manipulation, analysis, visualization, and modeling skills. Discuss challenges faced, methodologies used, and outcomes achieved.

- *Stay updated*: Be aware of recent advancements and trends in data science, ML, and AI. Understand how these advancements impact the field and how they can be applied in practical scenarios.

- *Mock interviews*: Practice technical interviews with peers or mentors. Simulate data science interview scenarios to get accustomed to articulating technical solutions and explaining your approach clearly.

- *Ask questions*: During the interview, don't hesitate to ask for clarification on questions or discuss different approaches. Communication of thought process is as important as the solution.

On top of this, coding questions are often a part of the technical interview process. You might be asked by an interviewer directly or given a coding exam. In the next section, we will provide suggestions on how to ace this section of the interview process.

Coding questions, step by step

Tackling technical coding questions in data science typically involves a structured approach to effectively solving problems. Here's a step-by-step framework:

1. *Understand the problem*:

    - Read the question thoroughly, making sure you understand the problem statement, input, and expected output.

    - If any part of the question is unclear, ask for clarification. It's crucial to have a clear understanding before proceeding.

2. *Define the approach*:

    - Identify the data requirements, including data structures or variables needed to solve the problem.

    - Choose the appropriate algorithms, data manipulation techniques, or models required to solve the problem efficiently.

3. *Design the solution*:

    - Outline the steps you'll take to solve the problem. This helps in organizing your thoughts before coding.

    - Think about boundary or edge cases that might affect your solution.

4. *Code implementation*:

    - Begin coding with simpler components or functions before tackling the entire problem.

    - Clearly document your code with comments to explain the logic and steps being implemented.

- Test your code with sample inputs, gradually increasing complexity to ensure it works as expected.

5. *Optimize and refactor*:

    - Analyze your code for areas where efficiency can be improved, such as reducing time complexity or optimizing memory usage.

    - Review, refine, and refactor your code to make it cleaner, more readable, and maintainable without compromising functionality.

6. *Communicate your solution*:

    - If in an interview setting, be prepared to articulate your thought process, explain the steps you took, and justify your choices.

    - Be open to suggestions or feedback on your solution and be ready to adapt or improve based on discussions.

7. *Review and learn*:

    - If errors occur, analyze why they happened and learn from them.

    - Review alternative solutions or best practices for similar problems to enhance your problem-solving skills.

This structured framework helps break down complex coding problems in data science into manageable steps, ensuring a systematic approach to problem-solving and coding efficiency.

In summary, mastering a technical data science interview involves a strong foundation in core concepts, practical application through projects, continuous practice in coding and problem-solving, and staying updated with the evolving landscape of data science and ML.

Assessment

Consider the following common data science problem: *Calculate the mean of a list of numbers while ignoring outliers. Apply the previous answer framework to solve this problem in Python.*

Answer

Here is how to apply the answer framework:

1. *Understand the problem*: Here, we want to calculate the mean of a list, excluding outliers. Often, outliers lay beyond the 10th and 90th percentiles.

2. *Define the approach*: To solve the problem, we need a list of numbers and a method to identify outliers based on percentiles. Plus, we'll use Python's NumPy library for calculating percentiles and statistics for mean calculation.

3. *Design the solution*: The pseudocode should do the following:

    - Calculate the 10th and 90th percentiles
    - Filter values falling within this range
    - Calculate the mean of the filtered values

4. *Code implementation*: Here is how we would implement the code:

```
import numpy as np
from statistics import mean

def calculate_mean_without_outliers(nums):
    lower_bound = np.percentile(nums, 10)
    upper_bound = np.percentile(nums, 90)

    filtered_values = [num for num in nums if lower_bound <= num
<= upper_bound]

    return mean(filtered_values)

# Test the function
data = [12, 15, 17, 19, 20, 21, 23, 25, 1000]  # Example list
with an outlier (1000)
result = calculate_mean_without_outliers(data)
print('Mean without outliers:', result)
```

5. *Optimize and refactor*: This code provides a straightforward solution. However, we might identify outliers using the **interquartile range (IQR)** method or, for larger datasets, consider optimizing the filtering process or exploring more efficient ways to identify outliers. For example, you might use some of the functions from the `sklearn` package such as the `IsolationForest` function – we have not covered this function in the book, but it is used to identify outliers in a dataset.

6. *Communicate your solution*: In an interview setting, explain the logic behind using percentiles to identify outliers and how the mean is calculated after filtering the data. For example, you might state, "The code sample will filter out the lowest and highest values in the dataset. Assuming there are outliers in the dataset, they will likely be filtered out. However, if given more time, another approach would involve first exploring the dataset with something such as a boxplot to identify outliers in the data. Additionally, we could use the IQR method to identify outliers. Once the code has removed the outliers from the dataset, it then computes the mean of the remaining values."

7. *Review and learn*: Reflect on the code, check for potential improvements or alternative methods, and learn from different approaches to solve similar problems. As stated before, we could improve upon our initial code by using the IQR method to identify if there are outliers in the dataset (our initial approach does assume that there are outliers):

```
import numpy as np

data = [12, 15, 17, 19, 20, 21, 23, 25, 1000]  # Example list
with an outlier (1000)
# Calculate Q1 and Q3
Q1 = np.percentile(data, 25)
Q3 = np.percentile(data, 75)
# Calculate IQR
IQR = Q3 - Q1
# Define lower and upper bounds
lower_bound = Q1 - 1.5 * IQR
upper_bound = Q3 + 1.5 * IQR
# Remove outliers
filtered_values = [num for num in nums if x >= lower_bound and x
<= upper_bound]

# Calculate the mean of the data without outliers
result = np.mean(filtered_values)

print('Mean without outliers:', result)
```

The panel stage

Encountering the **panel stage** signifies further advancement in the hiring process and suggests several key assumptions about your candidacy:

- *Cultural alignment*: Reaching this stage likely means that you've demonstrated a strong alignment with the company's culture and values. The panel may focus on assessing how well your personality and work style match the team and company ethos.

- *Competitive candidate*: Being interviewed by a panel suggests that you're among the top contenders for the position. You've likely stood out from other applicants and are being evaluated more comprehensively.

- *Comprehensive evaluation*: The interview panel stage often involves a comprehensive assessment of your skills, experience, and fit for the role – at this stage, it's likely they already think you'll be a good fit. Different panel members might focus on specific aspects relevant to their expertise or department.

During a panel interview, candidates can anticipate the following:

- *Diverse perspectives*: The panel may consist of individuals from various departments or levels within the organization. Questions may vary based on each panel member's area of interest or expertise.

- *In-depth technical and behavioral questions*: Expect a mix of technical questions related to the role, behavioral inquiries exploring past experiences, and situational scenarios to assess problem-solving skills.

- *Assessment of cultural fit*: The panel might explore how your values, working style, and personality align with the team and the company culture.

- *Team dynamics*: You might be evaluated on how well you could collaborate and contribute within the team. Panel members may observe how you interact with different personalities and respond to group dynamics.

- *Final evaluation*: If you have not already encountered a final evaluation assessment, you will likely encounter it at this stage. You may be asked to talk with a wider audience, including the members of the interview panel and potentially the hiring manager. This is typically the final straw before making a hiring decision. Here, the panel collectively evaluates whether you are the best fit for the role and the organization.

You should prepare for a more comprehensive evaluation during the interview panel stage, showcasing your skills, adaptability, and collaborative abilities and how they align with both the role and broader organizational objectives.

Summary

Approaching data science interviews involves a holistic preparation strategy tailored to different stages of the hiring process. Initially, at the recruiter stage, focus on crafting a precise, impactful resume highlighting relevant skills, projects, and experiences.

As you progress to the hiring manager stage, dive deeper into showcasing your alignment with the company culture, mission, and your ability to solve problems effectively. Engage in open discussions, highlighting your achievements and demonstrating enthusiasm for the role and organization. When facing the interview panel, emphasize adaptability and collaborative skills, engaging with diverse perspectives and showcasing your ability to integrate into varied team dynamics.

Lastly, during technical stages, emphasize a strong foundation in core concepts, practice problem-solving and coding, and stay updated with the latest trends in data science. Emphasize your ability to tackle complex problems methodically, communicate your approach clearly, and be open to feedback throughout the process. Tailoring your preparation to these distinct stages can significantly enhance your performance and chances of success in data science interviews.

At this point in the book, we will assume that you have done phenomenally during your interview process and the organization is looking to hire you. What comes next? Well, in the next chapter, we will dive into the topic of negotiation for things such as salary and benefits.

References

- [1] *40 Important Job Interview Statistics [2023]: How Many Interviews Before Job Offer* from *Zippia*: `https://www.zippia.com/advice/job-interview-statistics/`

14
Negotiating Compensation

In your journey toward your next role as a data scientist, the negotiation phase stands as the crescendo – the culmination of your efforts, skills, and worth. It's the moment where the dance of give and take begins, where your value converges with the company's offerings. This chapter serves as your compass through this pivotal phase, navigating the intricate terrain of negotiating compensation with HR.

From the tangibles, such as salary and stock options, to the intangibles, such as flexible hours and professional development perks, we'll delve into the spectrum of negotiables and equip you with the strategies to navigate this negotiation effectively. Join us as we unravel the art and science of securing not just the job, but a compensation package that echoes your true worth as a data scientist.

In this chapter, we will review the following:

- Understanding the compensation landscape
- Negotiating the offer

Understanding the compensation landscape

Congratulations, you've received a job offer! However, before embarking on the negotiation journey, it's crucial to map the compensation terrain. This entails delving into the company culture, industry norms, and the specifics of the job market.

To effectively research the appropriate salary range, consider not only the role's regional market value but also how your personal qualifications align with industry expectations. Resources such as Glassdoor, Payscale, Salary.com, and the **Bureau of Labor Statistics** (**BLS**) are invaluable in this process. They provide detailed salary benchmarks, considering factors such as location, years of experience, and the unique skill sets that the job demands. Utilize these platforms along with industry reports and networking connections to gather a comprehensive view of what competitors are offering for similar positions.

Why is this extensive preliminary research crucial? It sets the foundation for realistic expectations and informed negotiations. Consider the analogy of negotiating the price of a used car. Resources such as Kelley Blue Book offer a guide to reasonable price ranges based on specific criteria such as make, model, year, and mileage. Similarly, when negotiating compensation for a data science role, understanding the interplay between the job title, region, qualifications, and years of experience is key to estimating a fair salary range.

But it's not just about the salary. Understanding the full spectrum of compensation, including non-monetary benefits such as tuition reimbursement, flexible work hours, or work-from-home options, is equally important. This knowledge empowers you to negotiate a package that aligns with your career goals and personal needs, ensuring a fair and satisfying job offer. By comprehensively understanding the job landscape, you can effectively navigate the negotiation process, achieving a balance between personal value and market standards.

Negotiating the offer

Undertaking the journey to secure your ideal job offer requires more than just accepting the first proposal that comes your way. It's about understanding your worth, articulating your unique value, and strategically negotiating for what truly matters to you. In this section, we delve into the art of negotiation, guiding you through the essential considerations, from assessing your market value to understanding the full spectrum of the offer's elements. We'll explore various scenarios, including the nuances faced by new graduates and experienced professionals when pivoting careers, to demonstrate how tailored negotiation strategies can significantly impact your job offer.

Negotiation considerations

The first step in negotiating an offer is understanding your market value. This isn't always obvious, but it's important that this assessment is objective. To begin, highlight your unique skills, experiences, and accomplishments. Consider your contributions in previous roles and how they align with the new opportunity. How many of the job's requirements do you meet? Do you exceed any of the requirements? What about the preferred qualifications? The answer to these questions will help you objectively assess your value in the market for the particular role that you're entertaining.

Next, review your job search priorities. For example, rank your job preferences and identify non-negotiables. Will the job require a lot of overtime or travel? Does it require working in the office? Is there an expectation to reply to emails after hours? Is there a transparent promotion roadmap? How is the commute? It's important to understand these details and their significance to you.

When thinking about your priorities, you can organize different qualities of your job offer into two buckets, personal and material:

- *Personal*: These are the qualities of the job that bring you personal benefits. These might include the following:

  - Professional or career development

  - Advancement opportunities

  - Work experience fulfillment

  - Location

  - The person you'll report to

  - Mentorship

  - Flexible hours and/or work format (e.g., on-site versus remote or hybrid)

  - Interest in the business or industry (e.g., gaming, healthcare, education, etc.)

  - Travel opportunities

- *Material*: These are the qualities most people think of when considering compensation negotiations. These include the following:

  - Base salary

  - Bonuses

  - Stock options/equity

  - Benefits (e.g., health, dental, vision, and additional discounts and perks)

  - Retirement plans

  - Tuition assistance

  - Training

  - Paid time off (PTO) and vacation time

  - Relocation assistance

  - Health and wellness programs (e.g., gym memberships, mental health services, child services, etc.)

  - Office amenities

  - Company equipment (e.g., car, cellphone, laptop, etc.)

Companies may not be willing or flexible in negotiating all of the factors listed in the preceding lists, such as the person you'll report to in the role. However, understanding them will help you compare job opportunities and identify focus areas for your negotiations. You should focus your negotiations on the factors that you feel are must-haves and any areas in which you think the job offer is lacking compared to the current market standards you've identified in your research. For example, if tuition assistance is a must-have for you because you are planning on returning to school, you should include this as a part of your negotiations if the job offer does not include it.

Responding to the offer

After identifying your market value, and your personal and material preferences, it's time to negotiate. Most companies will give you about a week to consider the offer. However, there are instances where they need a response sooner. Alternatively, some companies are more lenient with the offer acceptance (or rejection) timeline because they're more concerned with your certainty about the role than their need to fill it. In either case, it is your job to make sure you and HR are on the same page. The deadline to accept the offer should be clear and, ideally, in written form. Be sure to request as much time as possible, particularly if you need it to discuss the details with loved ones or peers, or if you're anticipating other offers.

Despite having time to think over the details of the offer, you should negotiate the details ASAP. This may happen over the phone, video chat, or email. In reality, these negotiations last minutes – nothing more. Regardless of the format, thank the representative for relaying the good news and express your genuine excitement for the opportunity. After all, they've selected you over countless others, and this is something worth celebrating.

Then, it's time to lay your cards on the table. When doing so, it's best to be brief, concise, and confident. You've done your homework, so there's nothing to worry about. You could say something like this:

"I'm thrilled about the opportunity to contribute to the team and am very appreciative of the offer. However, based on my research and understanding of the market value for this role, as well as my specific skills and experiences, I would like to discuss the possibility of a salary that better reflects these factors. I believe a figure of [X amount] would be more aligned with the industry standards for someone with my qualifications."

In most cases, the HR representative will relay the request back to the hiring manager and follow up accordingly. The worst-case scenario is that they will reject your counteroffer. However, in a lot of cases, the offer will be amended to a **final offer**. A final offer means that there is no additional negotiation. At this point, you either "take it or leave it."

> **Note**
>
> One of the biggest and most frequent mistakes of job seekers is failing to negotiate at all. This is often due to 1) candidates undervaluing themselves, 2) a lack of knowledge of the role's pay range, and/or 3) the fear of rejection. While the first two reasons can be attributed to ignorance about the market or self-doubt, the last reason is almost irrational. It is very rare for jobs to retract an offer if the counter is reasonable and requested in a professional manner. Thus, whatever you do, negotiate!

Maximum negotiable compensation and situational value

Negotiating a salary is more of an art form than a science. There are countless articles, editorials, webinars, and even books, all designed to teach you how to advocate for yourself. Some popular books include the following:

- *You Are a Badass* by Jen Sincero (*Running Press Adult*)

- *Quiet: The Power of Introverts in a World That Can't Stop Talking* by Susan Cain (*Crown*)

- *Getting to Yes with Yourself (and Other Worthy Opponents)* by William Ury (*HarperOne*)

The effectiveness of negotiation strategies varies based on individual career goals, experience, knowledge of the job/company, and prevailing market conditions. Central to all of these approaches, though, is the concept of **situational value** – your unique contribution to a role, influenced by both internal and external factors. Your situational value is a combination of your experiences, unique skills, and personal attributes. Understanding and articulately conveying this value is key to maximizing the benefits you can negotiate in your job offer.

Figure 14.1 illustrates the importance of internal and external factors as well as your situational value in negotiating the maximum compensation:

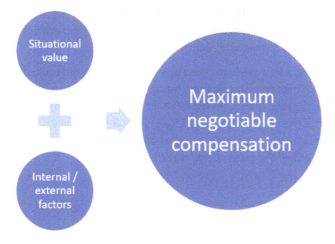

Figure 14.1: The maximum negotiable compensation equation

Let's break down the concept further for clarity. The internal factors that influence **maximum negotiable compensation**, which is simply the maximum compensation, material and personal, that you can negotiate, typically include the budget allocated for the role and the salary bands set for the position. These are elements internal to the company's structure and policies. On the other hand, external factors are those outside of the company's control, such as the regional salary range for the role or the current market demand for your skills.

Your situational value – comprising your specific experiences and skills and how they align with the role – interacts with these internal and external factors. For instance, even if a role has a high budget (an internal factor) and is in a high-demand field (an external factor), your ability to negotiate a better offer may be limited if your situational value doesn't align closely with the job requirements. Conversely, a strong situational value might not yield a significant salary increase if it exceeds the role's regional salary range or the company's budgetary constraints.

In the following sections, we'll explore examples that illustrate how this framework operates from various perspectives.

The college graduate

Suppose you are a new graduate seeking your first full-time data science role. You're new to the job market, but your experience is limited to an internship and school projects. Thus, your lack of experience will be a hurdle to overcome, especially as you compete against other grads with similar credentials. However, you've done your research!

Let's explore this scenario:

- *The job*: You're offered a full-time junior data scientist job for a gaming start-up. The company has multiple offices throughout the nation.

- *The offer*: The salary comes back much lower than expected. They're offering the lowest end of the market salary range, and the role is not bonus-eligible given its junior status. Additionally, there is no formal retirement program. The healthcare package is tolerable because you're young with no dependents or managed ailments.

- *Situational value*: As a new graduate, your negotiation options are limited. There isn't a wealth of full-time experience you can speak to, but you have a passion for gaming, an industry in which you hope to grow. In fact, you're very familiar with the company's game portfolio and have a respectable grasp on industry challenges and solutions that could improve its products and marketing plan. You even have a "proof of concept" project that you worked on in college where you've identified the cause of player churn.

- *The counter*: On a call, you express how thrilled you are about the opportunity and that you enjoyed meeting the team. Switching gears, you inform them that you understand the offer is on the lower end. However, you believe that your candidacy is uniquely advantageous, given your knowledge of the industry. You already have some ideas of how to increase player return rates and have worked on a similar project in the past. As the company is a start-up, you're aware that funding for this role is mostly set in stone. However, you're confident in the company, its products, and your pitch, so you request equity instead. You also request a company laptop with the necessary computational resources, and the ability to work from home.

- *The final offer*: Because you are fresh out of college, the employer doesn't feel comfortable with you working from home. They want you to have the opportunity to engage with the team in person and learn more about the company. However, they are willing to allow a hybrid model where you work from home two days a week. Furthermore, they're allowing you to choose your location of preference. You discover that the pay doesn't change based on the location, so you opt for their Columbus, OH location over their San Francisco, CA office. They also realize that providing you with a laptop is in their best interest to protect company data and to provide the best tools for the job. Lastly, they grant you 100 shares of company stock with a bi-annual re-evaluation period to grant additional stock based on performance. You happily take the opportunity, knowing that the role has growth potential and will provide excellent experience in the gaming industry.

In this negotiation scenario, you demonstrated several key strengths that contributed to a successful outcome. Firstly, your approach was marked by a combination of enthusiasm and pragmatism. By expressing excitement about the opportunity and acknowledging the offer's limitations, you struck a balance between eagerness and realistic expectations. Your strategic use of your unique value proposition, highlighting your passion for gaming and specific industry insights, effectively showcased how your skills and interests aligned with the company's needs. This not only underscored your potential contribution to the company but also provided a solid foundation for your negotiation requests.

Additionally, your understanding of the start-up's financial constraints led to a creative negotiation tactic, focusing on equity and practical benefits such as a company laptop and a flexible work location, rather than just salary. This adaptability and foresight to request equity and other non-monetary compensations demonstrated a keen understanding of the start-up environment and long-term career growth, ultimately leading to a mutually beneficial agreement.

The pivoter

In this scenario, you're a boot camp graduate with years of full-time experience, but not in data science. Instead, you've spent most of your career as a Paid Search Manager for a marketing firm. After graduating from the boot camp, you're looking to pivot to a data science role.

Let's explore this scenario:

- *The job*: You're offered a data scientist job on a new marketing sciences team at a digital marketing firm. The department has only just begun hiring data scientists, so you will only be the second hire, although it has plans to hire more. The interview panel was very impressed with your familiarity with digital marketing KPIs and strategies to improve paid search campaigns. You're also very familiar with some of the firm's clientele and, in turn, common challenges that those brands face in their respective industries. The role is hybrid, which is exactly what you prefer.

- *The offer*: HR sends you an email with an offer letter. You've got the job! The salary is what you expected – it's on the lower side of the job's regional range and just about where you expected to land, given the opportunity to grow in the role. The company offers pretty standard retirement and medical plans, but it doesn't mention anything about tuition assistance or professional development. This is important to you as a new data scientist with only a few months of boot camp knowledge. This is especially true since you're an early hire on the team. You also know that there's a huge opportunity to apply newly learned techniques to the job, such as applying neural networks or generative AI APIs to projects. The offer includes a host of discounts associated with the brands the company advises.

- *Situational value*: Due to the verbal exchanges during your interviews, you've noticed that many interviewers on the panel were impressed with your past experience as a Paid Search Manager. They were also excited to hear about some of the previous projects that you've worked on where you applied your knowledge of text mining to extract campaign insights and automate keyword generation. You know that not having full-time experience as a data scientist is the biggest gap you have to close in their minds.

- *The counter*: You reply to the email, thanking everyone involved for having faith in your candidacy. You're legitimately excited to get started and can't wait to join the team. Because you know there's room to negotiate your wage, you ask for 10% more than the original offer. You also state that you look forward to growing with the company and using analytics to solve tough business challenges. As a result, to remain current on the latest methods and ensure the most innovative solutions, you inquire about continued learning funding (such as tuition assistance or financial support for certifications). The HR representative says that they will check with the hiring manager and provide an update within 24 hours.

- *The final offer*: HR responds the following morning, and it's great news! Not only did the hiring manager grant your 10% salary increase request, but they also agreed that career development for your role is important. As a result, they are willing to pay up to $5,000 a year for any relevant terminal degree or certification program of your liking. They also mention that the previous hire (the senior data scientist on the team) has a lot of great experience but lacks some familiarity with digital marketing. As a result, they support a mentorship engagement between you both, which is a perk that will undoubtedly contribute to your growth as a data scientist!

In this negotiation scenario, you adeptly leveraged your unique background and keen understanding of the company's needs to secure a beneficial offer. The strategic move to highlight your past experience as a Paid Search Manager and the ability to apply this expertise in a new data science role was particularly effective. This approach not only demonstrated your value to the company but also addressed the gap in your data science experience.

By expressing genuine enthusiasm for the role and team, you fostered a positive tone for the negotiation. Additionally, your request for a salary increase was anchored with a reasonable percentage, reflecting an understanding of your worth and the market standards.

Your foresight in requesting support for continued learning and professional development was a smart move, emphasizing your commitment to growth and innovation in the role. This not only benefited you personally but also aligned with the company's interest in keeping its team updated with the latest industry practices. The successful negotiation of a mentorship with the senior data scientist further underscored your strategic approach to your career development, ensuring a comprehensive package that went beyond just financial compensation.

The grower

In this example, you're an experienced data scientist with a professional background that spans five years. You're currently employed, but you're seeking a more challenging opportunity that will stretch your skill set. You've mostly worked in the supply chain field and feel that you have a good chance of landing another similar role, considering you tend to interview well. You're leaving your current employer as there isn't a lot of growth opportunity, and despite your efforts and the praise you've received, they have not granted a raise in the last few years. You're now below the market range for your expertise, and you're seeking an opportunity to grow into management.

Let's explore this scenario:

- *The job*: You're offered a senior data scientist job at a pharmaceutical company in the supply chain department. You have very little experience in pharma, but you're practically a supply chain expert! The hiring manager believes that your previous experience is transferable and has faith in your ability to learn about the new industry environment, given the proper guidance.

- *The offer*: The interview process goes pretty smoothly, and you receive a call from HR confirming the company's interest in hiring you. It offers you the higher end of the expected salary range and a nice sign-on bonus. You're also granted company stock based on performance. You're fairly pleased with the material offer, but it's still unclear whether you'll be stuck in the role long-term or whether there's a formal process that leads to management.

- *Situational value*: You already know that you're a seasoned supply chain expert with tons of experience in machine learning. And despite your lack of experience in pharma, you know that you will be one of the more senior data scientists on the team. Not only were you explicitly told this, but you were able to confirm it by researching the existing team on LinkedIn. You take it upon yourself to reach out to a few of them to learn more about their experiences with the company. Your takeaway from the conversations is that the manager is great, and the projects are interesting, although you may be expected to put in some late hours from time to time. Additionally, there are some data hygiene issues that they're currently working through. You also notice that one of your former colleagues on LinkedIn used to work for the same company. You reach out to them to ask about their experience with the company, and they confirm that it's a great place. They were even willing to share their salary with you at the time of employment, which was 2 years ago. You learn that they were paid almost 15% more than your offer for the same role and level.

- *The counter*: To initiate the negotiation process, you request to speak with the hiring manager when they have a spare 15 minutes. HR arranges the call, and you briefly greet the manager. You're mostly satisfied with the material compensation, but you never settle for the original offer, and you know there's room for an increase, so you request a 15% increase. The request is supported by the information that you gathered from your former colleague. You suspect that you are the most senior data scientist on the team, so you feel comfortable asking for this increase. You also pitch your potential to really shape the team and identify some opportunities for improvement with processes, data quality, and governance. You support these details with examples from previous roles you've held and even point out a few pain points that you noticed during the interview. Before the call ends, you inquire about growth opportunities as you're interested in leading teams in the future.

- *The final offer*: After a few days, the hiring manager calls you directly. They're overall impressed with your skills, eagerness, and transparency. The hiring manager agrees to a 10% increase in salary, with the anticipation of reassessing your performance in 12 months. If you can demonstrate your ability to learn quickly (particularly details specific to the pharmaceutical industry), they are open to discussing a formal promotion to a managerial role. With the promotion would come a considerable salary increase. You're impressed with the partial wage increase and efforts to fulfill your interest in management, so you accept the role.

In this negotiation scenario, you demonstrated an exceptional blend of strategy, research, and communication skills. Your proactive approach in reaching out to current and former employees of the company provided valuable insights into the company culture, expectations, and salary benchmarks. This level of research not only equipped you with a realistic view of the role but also offered a solid foundation for your salary negotiation. By leveraging the salary information obtained from a former colleague, you were able to confidently gain a 10% salary increase, a move that showcased your awareness of market standards and your self-worth. Your approach to the negotiation was also characterized by a clear articulation of your value proposition.

You effectively highlighted your expertise in supply chain management and machine learning, and identified specific areas where you could contribute to process improvements and data governance. This not only underscored your suitability for the role but also demonstrated your potential for future leadership.

Additionally, your openness about your career aspirations and the request for a clear path to managerial roles displayed foresight and ambition. Your successful negotiation for a significant salary increase and a potential managerial promotion reflected your strong negotiation skills and strategic thinking, setting a positive tone for your future with the company.

Assessment

What were some situational values used to negotiate compensation in the previous negotiation examples?

Answer

- In the college graduate example, the candidate identified their enthusiasm for gaming as a situational value. Someone with that sort of passion may convince the hiring manager that they're there for the long haul and/or that they'll be genuinely interested in the work they'll be doing. They then explain their familiarity with the company's products and industry, which makes for a smoother onboarding experience.

- In the pivoter example, the candidate was able to speak to their years of relevant experience and they sold their eagerness to learn as a valuable investment for the company. The candidate also knew that there was room for a salary increase, given the initial offer, which was on the lower side of the regional salary range.

- In the grower example, the candidate recognized their position as a senior hire. They also conducted research on the salary range via an acquaintance and former colleague. In turn, they also pitched themselves as an excellent candidate for a managerial role under the correct guidance.

Summary

In the intricate dance of negotiating compensation, this chapter has unveiled a tapestry of strategies and insights crucial to orchestrating a successful negotiation. By understanding the multifaceted landscape of negotiables, from base salary to the nuances of work-life balance and career growth, you've equipped yourself with the tools to navigate this pivotal phase.

Emphasizing the significance of research, timing, and a strategic approach, you're poised to not only negotiate but to collaboratively craft a compensation package that reflects your true value. Situational value, that unique amalgamation of skills, experiences, and expertise you bring, becomes your guiding star in this negotiation journey. And within this negotiation lies the concept of maximum negotiable compensation. Your mastery of these negotiation tactics fosters an environment where both you and the employer find equilibrium, and a satisfying agreement that extends beyond a mere transaction, embodying a partnership founded on recognition of your worth as a data scientist. Remember, negotiation isn't just about securing a job; it's about securing the compensation that resonates with your values and aspirations in this ever-evolving landscape of data science.

Final words

As we draw the curtain on this journey toward cracking the data science interview, let us take a moment to reflect on the vast amount of knowledge and skills you have now acquired. From understanding the dynamic landscape of data science in *Chapter 1* to mastering the art of negotiation in *Chapter 14*, this book has been a comprehensive guide, aiming to sculpt you into a formidable candidate for your next data science role.

You have traversed the intricacies of technical interviews, delved deep into Python programming, SQL, machine learning, version control, and even explored the revolutionary realms of deep learning and MLOps. But beyond the technicalities, you have learned to present yourself, your skills, and your passion for data science in a way that resonates with recruiters and hiring managers. You have been equipped not just with knowledge, but also with the confidence to apply it effectively to real-world scenarios.

As you step into the job market, remember that each chapter of this book was a stepping stone toward your dream role. You are now well-prepared to not only face the challenges of job hunting and interviews but to excel in them. You have the tools to negotiate not just a job, but a career that aligns with your aspirations, values, and life balance. Furthermore, if you ever need a refresher on the main topics in the field, you can always come back to freshen up!

Thank you for allowing us to be a part of your journey. Your commitment and passion for learning is the true driving forces behind the progress you've made. As you embark on this exciting phase of your career, know that the wisdom, skills, and insights you have gained are your greatest allies. May your journey in data science be as fulfilling and impactful as the efforts you have put into preparing for it. Congratulations on reaching this milestone, and here's to the many successes that await you in the world of data science!

Index

Symbols

A

B

www.packtpub.com

Subscribe to our online digital library for full access to over 7,000 books and videos, as well as industry leading tools to help you plan your personal development and advance your career. For more information, please visit our website.

Why subscribe?

- Spend less time learning and more time coding with practical eBooks and Videos from over 4,000 industry professionals

- Improve your learning with Skill Plans built especially for you

- Get a free eBook or video every month

- Fully searchable for easy access to vital information

- Copy and paste, print, and bookmark content

Did you know that Packt offers eBook versions of every book published, with PDF and ePub files available? You can upgrade to the eBook version at packtpub.com and as a print book customer, you are entitled to a discount on the eBook copy. Get in touch with us at customercare@packtpub.com for more details.

At www.packtpub.com, you can also read a collection of free technical articles, sign up for a range of free newsletters, and receive exclusive discounts and offers on Packt books and eBooks.

Other Books You May Enjoy

If you enjoyed this book, you may be interested in these other books by Packt:

Data Science for Web3

Gabriela Castillo Areco

ISBN: 978-1-83763-754-6

- Understand the core components of blockchain transactions and blocks.
- Identify reliable sources of on-chain and off-chain data to build robust datasets.
- Understand key Web3 business questions and how data science can offer solutions.
- Build your skills to create and query NFT- and DeFi-specific datasets.
- Implement a machine learning toolbox with real-world use cases in the Web3 space.

Principles of Data Science

Sinan Ozdemir

ISBN: 978-1-83763-630-3

- Master the fundamentals steps of data science through practical examples.
- Bridge the gap between math and programming using advanced statistics and ML.
- Harness probability, calculus, and models for effective data control.
- Explore transformative modern ML with large language models.
- Evaluate ML success with impactful metrics and MLOps.
- Create compelling visuals that convey actionable insights.

Packt is searching for authors like you

If you're interested in becoming an author for Packt, please visit `authors.packtpub.com` and apply today. We have worked with thousands of developers and tech professionals, just like you, to help them share their insight with the global tech community. You can make a general application, apply for a specific hot topic that we are recruiting an author for, or submit your own idea.

Share Your Thoughts

Now you've finished *Cracking the Data Science Interview*, we'd love to hear your thoughts! Scan the QR code below to go straight to the Amazon review page for this book and share your feedback or leave a review on the site that you purchased it from.

https://packt.link/r/1-805-12050-6

Your review is important to us and the tech community and will help us make sure we're delivering excellent quality content.

Download a free PDF copy of this book

Thanks for purchasing this book!

Do you like to read on the go but are unable to carry your print books everywhere?

Is your eBook purchase not compatible with the device of your choice?

Don't worry, now with every Packt book you get a DRM-free PDF version of that book at no cost.

Read anywhere, any place, on any device. Search, copy, and paste code from your favorite technical books directly into your application.

The perks don't stop there, you can get exclusive access to discounts, newsletters, and great free content in your inbox daily

Follow these simple steps to get the benefits:

1. Scan the QR code or visit the link below

https://packt.link/free-ebook/978-1-80512-050-6

2. Submit your proof of purchase
3. That's it! We'll send your free PDF and other benefits to your email directly